DIALOGUE WITH THE MEDITERRANEAN

Dialogue with the Mediterranean
The Role of NATO's Mediterranean Initiative

Gareth M. Winrow

LONDON AND NEW YORK

First published 2000 by Garland Publishing, Inc.

2 Park Square, Milton Park, Abingdon, Oxon OX14 4RN
711 Third Avenue, New York, NY 10017, USA

First issued in paperback 2016

Routledge is an imprint of the Taylor & Francis Group, an informa business

Copyright © 2000 by Gareth M. Winrow

All rights reserved. No part of this book may be reprinted or reproduced or utilised in any form or by any electronic, mechanical, or other means, now known or hereafter invented, including photocopying and recording, or in any information storage or retrieval system, without permission in writing from the publishers.

Notices
Practitioners and researchers must always rely on their own experience and knowledge in evaluating and using any information, methods, compounds, or experiments described herein. In using such information or methods they should be mindful of their own safety and the safety of others, including parties for whom they have a professional responsibility.

Product or corporate names may be trademarks or registered trademarks, and are used only for identification and explanation without intent to infringe.

Library of Congress Cataloging-in-Publication Data is available from the Library of Congress.

ISBN-13: 978-0-8153-3624-2 (hbk)
ISBN-13: 978-1-1389-6763-2 (pbk)

To Marc Sinan and Nazan

Contents

Series Editor's Foreword		xi
Preface		xiii
List of Abbreviations		xvii

CHAPTER 1 The "Mediterranean": A "Region" of
             "Unstable Peace"?   1
           Introduction   1
           Is There a "Mediterranean Region"?   3
           The Mediterranean in the 1990s: An Overview   9
           The Middle East Peace Process   13
           The Need for Preventive Diplomacy   17
           Conclusion   19

CHAPTER 2 The Importance of "Dialogue" as a
             "Confidence-Building Measure"   23
           Introduction   23
           What Is Meant by "Security"?   24
           The Relevance of "Culture"   28
           The Importance of "Dialogue"   31
           CBMs and CSBMs for the Mediterranean   36
           Conclusion   44

CHAPTER 3 NATO and the Mediterranean in the Cold War   49
           Introduction   49
           NATO, the Mediterranean, and the Cold War   51
           NATO, Out-of-Area, and the Mediterranean   57
           Conclusion   65

CHAPTER 4	The Southern Mediterranean and the Middle East in the Post–Cold War Era: Political Developments	71
	Introduction	71
	Rulers and Elites	73
	Political Parties	76
	Religious Radicalism	79
	The Role of the Military	85
	Prospects for Democratization	87
	Conclusion	92
CHAPTER 5	The Southern Mediterranean and the Middle East in the Post–Cold War Era: Economic and Social Issues and Perceptions of Security	97
	Introduction	97
	The Problem of Migration	98
	The Current Economic Situation	103
	Trade Issues	106
	South–South Security Issues	108
	Southern Perceptions of Security	112
	Conclusion	116
CHAPTER 6	The Immediate Background to NATO's Mediterranean Initiative	121
	Introduction	121
	The Significance of Developments in the Gulf, the Balkans, and the Middle East	122
	Failed Trans-Mediterranean Cooperation Initiatives	131
	The Significance of NATO's Strategic Concept	139
	The Beginning of a Real Dialogue with the Mediterranean	144
	NATO Directs Its Attention toward the Mediterranean	154
	Conclusion	158
CHAPTER 7	NATO's Mediterranean Initiative	167
	Introduction	167
	Membership of the NATO-Mediterranean Dialogue	169
	The First Phase	172
	The Second Phase	175
	The Third Phase	178
	The Perspectives of NATO States	183

	The Perspectives of the Mediterranean Dialogue Countries	187
	Conclusion	191
CHAPTER 8	The Significance of Other Mediterranean Dialogues and NATO's Mediterranean Initiative	197
	Introduction	197
	The WEU-Mediterranean Dialogue and NATO's Mediterranean Initiative	199
	The OSCE-Mediterranean Dialogue and NATO's Mediterranean Initiative	205
	The Barcelona Process and NATO's Mediterranean Initiative	208
	Conclusion	217
CHAPTER 9	Conclusion	223
Bibliography		233
Index		247

Series Editor's Foreword

Gareth Winrow's new book confronts issues critical to the future of peace and prosperity in Europe and the Middle East. It is a pioneering work, dealing with an agenda of change along the political fault lines dividing the peoples of the "Euro-Mediterranean region." As Winrow cautions in the book's Conclusion, it is too early to speak of a sense of community among the states straddling the Mediterranean. But just as NATO's Mediterranean Initiative is a project in the creation of just such a community, so too is Winrow's analysis of it a big step in the scholarship of comtemporary relations between the European North and the largely Arab and Muslim South.

Events in Bosnia and Kosovo have recently demonstrated that it takes a good deal of professional courage to assess the meaning of the recent past and to speculate on the probabilities of the future. Good political science, however, does not shrink from this challenge, and Winrow's study of the NATO-Mediterranean dialogue tackles the complexities of relationships across cultural barriers with an aplomb that only a master of his material can muster. In the process Winrow has made a major contribution to the scholarship of the Atlantic Alliance as an institution-in-transformation. Since the Cold War a good deal of ink has flowed over change within the Alliance, but very little has been devoted to NATO's involvement in the development of civil society among the states of the Mediterranean beyond its crisis-management commitments in the former Yugoslavia. The latter has made the Alliance indispensible to the maintenance of peace in the Balkans, but at the same time NATO briefings are now routinely attended by delegations from Israel and the Arab states. In short, the present ambit of the Alliance's community-building activities in the Mediterranean is both broad and potentially much broader.

The Mediterranean Initiative is therefore integral to the *redefinition* of European security that began tentatively with the end of the Cold War and accelerated in response to the twin challenges of political reform and crisis on Europe's periphery during the 1990s. That redefinition involves the Alliance simultaneously in traditional "hard" security as well as in innovations in "soft" or "human" security. Winrow rightly stresses that NATO's strongest asset remains the combined military expertise of its members. That being the case, the Mediterranean dialogue will inevitably focus much of its attention on military cooperation and confidence-building initiatives between European and Arab governments. A history of mutual suspicion, however, means that efforts in military cooperation will be doomed unless NATO's diplomacy is especially sensitive both to domestic political realities within the Arab states and to the progress of relations between Israel and its neighbors, particularly the Palestinians. In contrast to NATO's experience with former Warsaw Pact states, the Alliance will therefore have to proceed with extreme caution on soft security issues such as human rights and civilian control over armed forces.

Advances will be slow and incremental. Above all, it will not be widely publicized. For that reason students of European and Middle Eastern affairs are indebted to Gareth Winrow for his superb scholarship on a subject that has hitherto attracted little of the attention it merits. The Mediterranean Initiative has enormous long-term implications for the security of Europe and the chances of peaceful coexistence in the Middle East. As series editor, I take great pride in adding *Dialogue with the Mediterranean,* an example of political science at its best, to Garland's collection on contemporary European issues.

Carl Cavanagh Hodge

Preface

Since this work was completed in mid-April 1999, a number of significant developments have taken place that may have some bearing on the security and stability of the Mediterranean. These, in turn, may have an impact on the future of NATO's Mediterranean Initiative.

The importance of the success of the Middle East peace process for the NATO-Mediterranean dialogue and other cooperative initiatives covering the Mediterranean is emphasized throughout this book. The defeat of Netanyahu and his government in the May 1999 elections in Israel raised hopes and expectations that further progress in the Middle East peace process would rapidly ensue. Earlier, Arafat and the PLO's Central Council were praised by many in the international community for opting not to declare unilaterally an independent Palestinian state on May 4, the date by which a permanent agreement between the Israelis and Palestinians should have been concluded. The PLO had decided to reconsider their position after assessing the results of the Israeli elections.

In the elections in Israel on May 17, 1999, Ehud Barak of the Labour party scored a landslide victory over Netanyahu for the post of prime minister. Barak secured over 56 percent of the vote. Netanyahu immediately announced his withdrawal for the time being from politics. Barak was handed the task of forming a new coalition government within forty-five days. Hopes were high that Barak would fulfill his pledge to withdraw Israeli troops from Lebanon within one year. Peace talks with Syria would need to be resumed in order to accomplish this withdrawal. Expectations were also raised that Barak would carry out the provisions of the Wye River Accord of fall 1998.

However, Barak first had to form a coalition government. This was no easy task given that Barak's One Israel umbrella group in the May 1999 elections had only won 26 seats in the 120-seat Knesset. In order to form a majority government, Barak was faced with the prospect of possibly including the Likud party within a broad-based coalition. Another option, equally problematic for Barak, was to include within the coalition the seventeen members of the Jewish ultraorthodox party Shas.

The final actions of the outgoing Likud-led administration created further complications. Much to the outrage of the Palestinians, a new wave of settlement-building on the West Bank commenced in late May/early June 1999. Most controversially, Israel's outgoing defense minister, Moshe Arens, approved plans to enlarge the Jewish settlement at Maale Adumim, which was situated between Jerusalem and Jericho. This would effectively extend Jerusalem's borders so far eastward that the West Bank would be split into two halves.

Partly to appease the increasingly frustrated Palestinians, Barak announced on June 7, 1999, that he would curtail the settlement-building of the outgoing government. The Labour leader also declared that he would establish a committee of cabinet ministers who would review settlement projects approved of by the Netanyahu administration. This committee would have the power to cancel any of these earlier projects.

What is clear is that a new Barak government will need to act quickly over the settlement issue in order to revive the stalled Middle East peace process. Other provisions of the Wye River Accord could then be tackled before dealing with much more problematic issues including the status of Jerusalem. The future course of development of NATO's Mediterranean Initiative and the EU's Barcelona Process and other cooperative initiatives covering the Mediterranean is closely tied in with what will happen in the Middle East over the coming months.

With reference to NATO in particular, the summit meeting in Washington, D.C., in late April 1999 resulted in a Summit Communiqué and the release of a new Strategic Concept for the Alliance. The aim of the latter was to take into account the new security environment of the post–Cold War era and to update the previous Strategic Concept, which had been announced in Rome in November 1991, at a time when the Soviet Union still existed.

A part of the new Strategic Concept focuses on themes outlined and discussed in 1991. However, taking into account NATO's involvement in Bosnia and Kosovo, extensive reference is made to crisis management. The importance of so-called peace support operations and crisis response operations is underlined. Other key issues such as the process of NATO

Preface

enlargement, the further development of the European Security and Defense Identity (ESDI), and the Alliance's relations with Russia and Ukraine are also addressed. In line with the original Strategic Concept, the April 1999 version also discusses the various risks and challenges to the security of NATO. Ethnic and religious rivalries, territorial disputes, failed attempts at reform, the abuse of human rights, terrorism, the disruption of the flow of vital resources, and so forth, could have an impact on individual NATO members or the Alliance as a whole. Concern over the proliferation of weapons of mass destruction and their means of delivery is also again emphasized.

Paragraphs 38 and 50 of the new Strategic Concept refer specifically to NATO's Mediterranean Initiative. In paragraph 38 there is the customary reference to how security in Europe is closely linked to the security and stability in the Mediterranean. The importance of the Mediterranean dialogue for NATO's cooperative approach to security and the Alliance's interest in confidence-building is stressed. The paragraph concludes:

> The Alliance is committed to developing progressively the political, civil and military aspect of the Dialogue with the aim of achieving closer cooperation with, and more active involvement by, countries that are partners in this Dialogue.

It is clear, then, that for the foreseeable future attention will be given to deepening but not widening the dialogue. Other non-NATO Mediterranean countries, such as Algeria, will not be admitted into the dialogue in the short to medium term. The preceding quotation also indicates that the Mediterranean dialogue countries are still not actively involved in the Mediterranean Initiative.

Paragraph 50 underlines the value of military-to-military contacts to deepen NATO's relationships with Russia, Ukraine, and the Mediterranean dialogue countries. It is noted that these confidence-building activities could be developed further. Reference is made to the possibility of consultation and cooperation in key areas such as, for example, training and exercises, civil-military relations, defense planning, crisis management, proliferation issues, and participation in operational planning and operations. It is not made clear in which of these specific areas in the military field NATO and the Mediterranean dialogue countries in particular could cooperate, although NATO's Mediterranean Initiative has certainly increasingly focused on building cooperation in the military and defense spheres.

Paragraph 29 of the Washington Summit Communiqué also refers to the importance of strengthening cooperation between NATO and the Mediterranean dialogue countries in the military field. Significantly, in close juxtaposition to the communiqué's paragraph on NATO and the Mediterranean, paragraphs 30 and 31 focus on the problem of the proliferation of weapons of mass destruction.

The possibility of NATO and the European Union (EU) more closely coordinating their activities in the Mediterranean may have been enhanced after the EU Summit meeting held in Cologne in early June 1999. It was decided that Javier Solana, the current NATO Secretary General, will be appointed as the first High Representative of the EU who will be responsible for the development of the EU's Common Foreign and Security Policy (CFSP). The Cologne meeting also appeared to confirm that an eventual merger between the Western European Union (WEU) and the EU will take place, probably by the end of 2000. The work of the WEU-Mediterranean dialogue could then become subsumed within the EU's Barcelona Process.

In spite of these developments the basic thrust of the arguments used in this book remains unchanged. Steady, but unspectacular progress within the NATO-Mediterranean dialogue is likely, assuming that the Middle East peace process will not collapse or suffer a serious setback under a Labour-led coalition government in Israel.

Research for this book was made possible by a NATO Research Fellowship. I am grateful to all diplomats and officials in Brussels whom I interviewed and who expressed their personal views and provided invaluable insights and information. Special thanks go to Catherine Guicherd at the North Atlantic Assembly and Professor Georges Delcoigne of the Université de Libre in Brussels for supplying documentation and establishing links with key individuals. I am also especially grateful to Emel Üresin, the Turkish Liaison Officer at NATO, who provided invaluable contacts. Ayşe Özakıncı, the Director of the American Information Resource Center in Istanbul was also very helpful. Particular thanks go to Professor Richard Gillespie of Portsmouth University for his support.

My thanks also go to those connected with Garland Publishers who assisted me in preparing this book in its final stages. Here, particular words of gratitude go to Carl Hodge and Mia Zamora.

Last but not least, special thanks go to my wife, Nazan, and my son, Marc Sinan, who patiently endured my many hours of absence and who always provided encouragement and support.

June 1999

List of Abbreviations

ACE	Allied Command Europe
ACRS	Working Group on Arms Control and Regional Security
AFSOUTH	Allied Forces Southern Europe
AFTA	Arab Free Trade Area
AIS	Islamic Salvation Army
AMU	Arab Maghreb Union
CBM	Confidence-building measure
CCMS	Committee on the Challenges of Modern Society
CENTO	Central Treaty Organization
CFSP	Common Foreign and Security Policy
CJTF	Combined Joint Task Force
CSBM	Confidence- and security-building measure
CSCE	Conference on Security and Cooperation in Europe
CSCM	Conference on Security and Cooperation in the Mediterranean
CSCME	Conference on Security and Cooperation in the Middle East
EAPC	Euro-Atlantic Partnership Council
EC	European Community
EEC	European Economic Community
EPC	European Political Cooperation
ESDI	European Security and Defense Identity
EU	European Union

EUROFOR	European Rapid Operational Force
EUROMARFOR	European Maritime Force
FFS	Front of Socialist Forces
FIS	Islamic Salvation Front
FLAM	Mauritanian African Liberation Forces
FLN	National Liberation Front
GIA	Armed Islamic Group
HASM	Islamic Movement of Mauritania
HATM	Movement of Reform and Renewal in Morocco
IAF	Islamic Action Front
IFOR	Implementation Force
IMF	International Monetary Fund
IPU	Inter-Parliamentary Union
JSRC	Joint Sub-Regional Command
MCG	Mediterranean Cooperation Group
MENA	Middle East–North Africa
MENAFIO	Middle East and North Africa Financial Intermediary Organization
MPS	Islamist Movement for a Peaceful Society
MSG	Mediterranean Special Group
MTI	Islamic Tendency Movement
NAA	North Atlantic Assembly
NAC	North Atlantic Council
NACC	North Atlantic Cooperation Council
NAVOCFORMED	Naval On-Call Force Mediterranean
NDP	National Democratic Party
NGO	Nongovernmental organization
NPT	Nuclear Non-Proliferation Treaty
OAU	Organization of African Unity
OCME	Organization for Cooperation in the Middle East
OSCE	Organization for Security and Cooperation in Europe
PfP	Partnership for Peace
PI	Istiqlal Party
PLO	Palestine Liberation Organization
POLISARIO	Popular Front for the Liberation of Saguia el Hamra and Rio de Oro
PRDS	Republican Democratic and Social Party
RC	Regional Command
RCD (Algeria)	Rally for Culture and Democracy

List of Abbreviations

RCD (Tunisia)	Constitutional Democratic Rally
RDF	Rapid Deployment Force
REDWG	Working Group on Regional Economic Development
RMP	Renovated Mediterranean Program
RND	National Democratic Rally
SC	Strategic Command
SFOR	Stabilization Force
STANAVFORMED	Standby Naval Force Mediterranean
UC	Constitutional Union
UDF	Union of Democratic Forces
UN	United Nations
UNPROFOR	United Nations Protection Forces
US	United States
USFP	Socialist Union of Popular Forces
WEU	Western European Union
WMD	Weapons of mass destruction

DIALOGUE WITH THE MEDITERRANEAN

CHAPTER 1
The "Mediterranean": A "Region" of "Unstable Peace"?

INTRODUCTION

In February 1995 NATO announced the launching of its so-called Mediterranean Initiative. A dialogue commenced with what were referred to as five "Mediterranean nonmember countries," namely, Egypt, Israel, Mauritania, Morocco, and Tunisia. Before the end of the year another state, Jordan, was added to the dialogue. The addition of Jordan enhanced the Middle East component of NATO's Mediterranean Initiative. NATO officials later referred to these six states as *Mediterranean dialogue countries* or *Mediterranean dialogue partners*. On strict geographical grounds, however, Jordan and Mauritania are not a part of the Mediterranean. They do not have a Mediterranean coastline.

The general aim of the dialogue was to contribute to "security" and "stability" in the Mediterranean as a whole. NATO's Mediterranean Initiative has gradually evolved. As well as continuing discussions, various cooperative activities have also been introduced. In 1997 the NATO-Mediterranean dialogue was given a more visible political profile when NATO officials decided to establish a Mediterranean Cooperation Group (MCG). The MCG was responsible for the further development of the dialogue. Steps were then taken to include cooperative activities in the military field between NATO and the Mediterranean dialogue countries. In short, the Mediterranean Initiative appears to have become a permanent feature of the Atlantic Alliance.

Although the NATO-Mediterranean dialogue has gradually evolved and assumed increasing importance, commentators, in general, have given it minimal attention. Rather, other issues involving NATO have been the focus of concern. Much has been written on the prospects and consequences of NATO enlargement and the future of NATO-Russian relations. The role of NATO within a so-called European Security and Defense Identity (ESDI), and the debates with regard to how to reform and adapt NATO's command structures and forces to make them more relevant to the post–Cold War era, are other subjects that have been discussed in detail. Concerning the southern and eastern Mediterranean, territories that were referred to in the past in NATO circles as "out-of-area," commentators and analysts have preferred to examine the role of the so-called Barcelona Process initiated by the European Union. Certainly, the Barcelona Process may be able to achieve much in the economic and social, and even in the political and military fields, in the Mediterranean area. However, NATO's Mediterranean Initiative should not be overlooked. The NATO-Mediterranean dialogue is a useful tool of preventive diplomacy and an important confidence-building measure (CBM). It may complement the activities of other institutions and forums interested in the Mediterranean, such as the EU, the Organization for Security and Cooperation in Europe (OSCE), and the Western European Union (WEU).

There are areas of potential and actual instability in the southern and eastern Mediterranean that may seriously affect the security of individual NATO members and the Atlantic Alliance as a whole. A number of NATO states in southern and eastern Europe form as it were a front line against risks and challenges and possibly even threats stemming from neighbors to their immediate south and east. France, Italy, Portugal, and Spain are particularly worried about political, social, and economic problems in North Africa. The Turkish authorities are concerned about developments in Iran, Iraq, and Syria, which may have serious repercussions for the Kurdish issue within Turkey. The Israeli-Palestinian problem appears to be far from being resolved in spite of the beginning of a Middle East peace process. The United States (US) administration, with its traditional close ties to Israel, and its geostrategic and geoeconomic interests in the Middle East and in the Gulf, is especially eager to promote the further strengthening of the Middle East peace process. The European members of NATO have also attempted to influence positively events in the Middle East. In the post–Cold War era, issues involving the security of the Middle East/eastern Mediterranean (and, also, the Gulf region) affect the security and stability of states to their west in the southern Mediterranean.

It is important first to have a clear impression of what one means when referring to the "Mediterranean." This chapter will address this question and will then provide an overview of the situation in the Mediterranean in the 1990s from the "security" angle. Throughout this study the crucial significance of the Middle East peace process for the future success of the NATO-Mediterranean dialogue and other dialogues and initiatives that are concerned with the Mediterranean are emphasized. Therefore, this chapter will discuss, in brief, the problems that the Middle East peace process has encountered since the formation of the Likud-led coalition government in Israel in May 1996. Obviously, though, the Middle East peace process is a huge subject that cannot be comprehensively analyzed in a few paragraphs. The need for preventive diplomacy in the Mediterranean is then examined. This will lead to a discussion of the role of CBMs and of dialogue, in particular, in the Mediterranean, which will be the focus of the second chapter.

IS THERE A "MEDITERRANEAN REGION"?

What does one mean by the "Mediterranean"? What geographical area should be included? Should other criteria be used in addition to geography to explain what one means by the Mediterranean? Should the so-called Middle East—or, at least, a part of it—be included in the Mediterranean? It is far from clear what area the Mediterranean encompasses. There is also a debate over whether the Mediterranean, however it is depicted, may be regarded as a specific "region." This begs the question, what is a region?

As already noted, the Mediterranean may be defined in strict geographic terms as consisting of the territories of only those states that have a Mediterranean coastline. This would exclude Jordan and Mauritania, two of the Mediterranean dialogue countries in NATO's Mediterranean Initiative. It has been observed that although the Mediterranean may be regarded as a "geographic entity," this did not make the Mediterranean "a political or strategic whole."[1] States that do not have a Mediterranean coastline have been incorporated in so-called Mediterranean cooperative schemes, in practice. The membership of the Conference on Security and Cooperation in the Mediterranean (CSCM) proposed by Spain and Italy in 1990 would have included states stretching from Mauritania to Pakistan.

According to Roberto Aliboni, in a monograph published in early 1991, in political and geopolitical terms, the Mediterranean covered states of southern Europe, the Maghreb, parts of the Arabized Sahel such as Chad, the Arab Orient, the Gulf countries, and the Horn of Africa, but not the Balkans. The Balkans was omitted on the grounds that it was a "crisis spot" that was distinct from the area to the south of Europe. In Aliboni's opinion it would be difficult to imagine a connection between crises in the Balkans and tensions in the southern Mediterranean, as they would stem from different political and security issues.[2] However, in the Bosnian crisis at the time of the violent disintegration of the former federal Yugoslavia, many Muslims in North Africa and in the Middle East complained at what they perceived to be Western inaction in the Balkans. Several Arab and Muslim states later participated in the international peacekeeping forces established after the cessation of hostilities in Bosnia—the Implementation Force (IFOR) and, its successor, the Stabilization Force (SFOR). There do seem to be links between events in the Balkans and in the southern and eastern Mediterranean that should not be overlooked.

In this study, the "Mediterranean" is examined with reference to NATO's Mediterranean Initiative. An attempt is made to understand how the Mediterranean is perceived by NATO officials and by the governments and publics of non-NATO states that are currently involved in the Initiative or that may be involved in the Initiative in the future. In this context, Mauritania and Jordan are therefore regarded as part of the Mediterranean. There are a few connections between the Balkans and the Mediterranean Initiative. However, the Balkans does have its own problems and so it will not be a focus of attention in this study.

According to NATO officials, therefore, there is in practice an overlap between the Mediterranean and the Middle East. Egypt, Israel, and Jordan are clearly part of the Middle East but they may also be regarded as eastern Mediterranean. It has been suggested that according to US strategic thinking, the Mediterranean is "the place where the Persian Gulf begins."[3] A linkage between events in the Gulf and in the southern and eastern Mediterranean was demonstrated by the extent of the popular reaction in the wake of the Iraqi invasion of Kuwait. It appears that one reason for Jordan's inclusion in the Mediterranean Initiative is Amman's potential positive influence in the Persian Gulf. However, it seems that there is a consensus in NATO circles that the Mediterranean Initiative itself should be perceived as a dialogue that does not intend to include the troubled Gulf in its mandate. There is a concern that the NATO-Mediterranean dialogue may be adversely affected by a crisis in the Gulf. In contrast, events in the Horn of Africa have no immediate bearing on the Mediterranean Initiative, while only France and the United States have expressed serious concern about past Libyan encroachments in Chad.

May one then speak of a Mediterranean region? According to Louis J. Cantori and Steven L. Spiegel, a region

> consists of one state, or two or more prominent and interacting states which have some common ethnic, linguistic, cultural, social and historical bonds, and whose sense of identity is sometimes increased by the actions and attitudes of states external to the system.[4]

According to this argument, therefore, a region may form part of one state or cover the territory of two or more presumably neighboring states. States that may form a region should share certain common features. The behavior of states located outside the region—"intrusive states," in the words of Cantori and Spiegel—may strengthen the sense of regional identity with regard to the governments and publics of states within the region. One may also add that regions are difficult to delineate. The boundary of one region may overlap with the boundary of another.

Cantori and Spiegel underlined the importance of the intensity of interaction between states or units within a region. These relations could be of a cooperative and/or antagonistic nature.[5] Commentators have also noted the significance of interaction between different units. They have defined regions in systemic terms, analyzing a region as a subsystem of the international system or viewing a region as a system itself. These analysts have also focused more on "security" issues in their explications of what is meant by a region. For example, W. Howard Wriggins has referred to "regional security systems" whose member states interact with each other so intensely that behavior in any one of them "is a necessary element in the calculation of others."[6] This depiction of a regional security system appears to be similar to Barry Buzan's definition of a security complex as "a group of states whose primary security concerns link together sufficiently closely that their national securities cannot realistically be considered apart from one another."[7] According to both Wriggins and Buzan, relations between units in their system/complex could be based on different degrees of amity or enmity: relations could be cooperative or antagonistic. In their emphasis on the importance of interaction between units, Wriggins and Buzan did not share the viewpoint of Cantori and Spiegel who had stated that the units should also share certain common features. There is no necessary connection between the level and nature of interaction and the degree of social and cultural homogeneity between the units in question.

According to one definition, regionalism consists of a set of policies pursued by one or more states (or units) within a region. These policies are aimed at promoting the emergence of a "cohesive regional unit" where cooperation between states (or units) across a wide range of issues is possible.[8] Here, there is an assumption that states have common interests. Regionalism does not exclude the possibility that states within a region may also have significant relations with states outside the region and with other regions.

In practice, regions may be evolving or "in making." They may be likened to "imagined communities" talked about and written into existence, as in the case of nations. The role of political actors and so-called regional builders is crucial. Moreover, like a nation, a region needs to be recognized and acknowledged by political actors outside of the region.[9]

A sense of regional identity may be shaped by the activities within an area of political elites in key states. Other groups such as businessmen, intellectuals, and nongovernmental organizations (NGOs) may help to promote common interests within an area. Touristic, educational, and cultural exchanges could also foster a regional identity.[10] Governments may not agree about what sort of regional arrangements should be established. Rival region-building projects may exist. The political elites of certain states may compete with one another to present themselves as the core or hub of regional cooperation.[11]

No region may be completely isolated and closed off from the rest of the world. Regions may differ, though, according to the degree of "autonomy" they may have. In this context, autonomy refers to "a state or region's ability to keep outsiders from defining the issues that constitute the local agenda."[12] As noted earlier, states beyond the region may positively affect the development of a regional identity. These intrusive states may encourage region building, perhaps in order to promote stability. Autonomy, in this instance, may not be beneficial for the strengthening of a regional identity. Or neighboring states may seek to expand their ties with one another to prevent intrusive states from interfering in their internal and external affairs. Intrusive states or other outside actors (e.g., alliances) may also negatively affect the development of a regional identity. This would likely occur when rival outside states or groupings of states seeking to enhance their geopolitical position are competing for the allegiance of the governments of states within a particular area.

In short, in their efforts to define a region, several commentators

The "Mediterranean": A "Region" of "Unstable Peace"? 7

have stressed the importance of geographic proximity and underlined that there should be a certain level of interaction between units within the region. This interaction may be cooperative and/or conflictual, though advocates of regionalism focus on positive interactions between states. There may be a degree of social and/or cultural homogeneity between the units in question but this is not essential. The role of outside actors should also be considered. Regions evolve and may develop. They could also in time, in effect, wither away if the interaction between units within the region substantially declines. This could perhaps occur due to a change in the international environment. Competition between outside actors to acquire client states or extend their spheres of influence could also result in a lessening of the sense of a regional identity.

May one then speak of a Mediterranean region? In his classic work, Fernand Braudel observed that four or five centuries ago the Mediterranean was a world in itself. The Mediterranean was a center of influence where peoples interacted through commercial exchanges and other forms of contact. States were in cooperation and conflict with one another.[13] Other analysts also refer to today's Mediterranean as a vital crossroads open to influences and exchanges, but they also tend to stress that the Mediterranean has become a "frontier" separating different worlds. According to one depiction, for example, a largely economically and politically developed Judaeo-Christian world to the north of the Mediterranean is separated from an Islamic world to the south that is economically undeveloped and ruled by authoritarian governments.[14]

According to Stephen C. Calleya, the Mediterranean "is a frontier separating different political, economic, military and cultural forces."[15] He strongly argues that there is no Mediterranean region. Rather, there are three subregions that cover the Mediterranean area, namely southern Europe, the Levant (eastern Mediterranean), and the Maghreb (southern Mediterranean). Southern Europe is a part of the region of western Europe, and the Levant and the Maghreb cover part of a separate Middle East region. Although there is some interaction between these three subregions, Calleya believes that they are becoming more embroiled in the concerns of their respective regions. As the disparities between the north and south multiply, the Mediterranean is becoming rapidly a "faultline between two separate and increasingly polarised regions."[16]

On the other hand, according to Fred Tanner, there is an emerging Mediterranean region that is increasingly acquiring its own distinct political identity. There is a trend of regionalization in the Mediterranean area, and this is evident in such initiatives as the Barcelona Process and the attempt to promote cultural and social cooperation through the Mediterranean Forum. In Tanner's opinion, there are political and economic problems in the Mediterranean. However, he believes that a common geography, history, and certain perceptions of togetherness are encouraging the emergence of a Mediterranean region.[17] It is worth noting, though, that texts and communiqués relating to the Barcelona Process refer to a "Euro-Mediterranean area" and not a Euro-Mediterranean region.

Certainly, in practice, there is an increasing interaction between businessmen, politicians, intellectuals, NGOs, and so forth, in the Mediterranean or Euro-Mediterranean area. It seems to be too early, though, to declare that a Mediterranean region is actually "in making," For example, the aborted attempt to establish a Conference on Security and Cooperation in the Mediterranean (CSCM) and the failure of Malta to inaugurate a Council of the Mediterranean based on the Council of Europe suggest that any trend toward regionalization in the Mediterranean is far from complete. Arguably, a future Mediterranean region would overlap with regions to its north (Europe), east (the Middle East), and south (the Sahel and Sub-Saharan Africa, the latter if Mauritania is included in a Mediterranean region). In spite of this overlap, a Mediterranean region could exist independently. For this to be the case, a Mediterranean region would need to have its own particular agenda, which would enable the region to have a measure of autonomy.

As already noted, cultural and social homogeneity between units in a region is not necessarily important. A number of commentators have rather placed emphasis on the intensity of interaction between units in order to substantiate whether or not a region may exist. It will be observed that there is a large economic and social disparity and a cultural divide between states north and south of the Mediterranean. But, with the end of the Cold War, there does appear to be an upsurge in contacts between states across the Mediterranean, although not to the extent that the actions of one state are always a necessary element in the calculations of other states in the area. It is important to note that the activities of politicians, businessmen, NGOs, and so on in the Mediterranean area, and the dialogues conducted by various initiatives such as those of the EU, WEU, OSCE, and NATO with regard to the Mediterranean, are not aimed primarily at creating a Mediterranean region. Rather, their objective is to strengthen cooperation and enhance "stability" and "security" in the Mediterranean area. One potential beneficial spin-off from these developments, however, is that the governments and peoples around the

Mediterranean may increasingly feel that they do share some sort of common identity, and this may ultimately encourage the pursuit of policies that aim to promote regionalism in the Mediterranean.

Obviously, NATO officials are not engaged in a debate about what is meant by the Mediterranean and whether a Mediterranean region exists or is "in making." They do refer to the Mediterranean as an area where developments may have an impact on the security of Europe, though the boundaries of this Mediterranean area are not defined. As already noted, the Mediterranean Initiative involves non-NATO states along the southern and eastern Mediterranean coastline together with other neighboring states in the Maghreb or Middle East. NATO member states that border the Mediterranean Sea to the north should also be included as a part of the Mediterranean even though they are also a part of Europe. Turkey is here included as a part of Europe at least with regard to security issues. As regions overlap, states may also be a part of more than one geographical area. The Mediterranean area—and possibly emerging region—as depicted by NATO, therefore, includes states that are a part of Europe, North Africa, and the Middle East. This is in line with the Euro-Mediterranean area as depicted by the EU's Barcelona Process. It will be seen that the role of the United States as an outside/intrusive actor in the Mediterranean has been, and is, of paramount importance. The United States is influential in setting many, but by no means all, of the issues that constitute the Mediterranean agenda. For example, the United States is not a full party to the EU's Barcelona Process, which is heavily involved in tackling various issues and problems in the Mediterranean.

It is worth noting here that Arab governments and even more so Arab peoples in the southern and eastern Mediterranean tend to regard NATO and the United States as one and the same. In practice, of course, NATO is not a monolithic organization totally dominated by the United States. The US administration does not necessarily share the same views as its European allies with reference to the Mediterranean. Not always succeeding in convincing its NATO partners to pursue a particular policy, the United States at times has also been prepared to accommodate the interests of those NATO member states especially concerned about developments in the southern and eastern Mediterranean.

THE MEDITERRANEAN IN THE 1990S: AN OVERVIEW

When discussing issues pertaining to the "security" and "stability" of the Mediterranean in the 1990s, various ongoing dynamics should be considered. There are problems in relations between states north of the Mediterranean. The interaction between states north and south of the Mediterranean must be taken into account. Relations between states south of the Mediterranean are also problematic. The security situation within particular states in the south—the intrastate as opposed to interstate dimension—should not be overlooked. The Mediterranean Initiative of NATO is primarily concerned with the north-south interstate interaction.

Concerning relations between states north of the Mediterranean, the crisis in the Balkans and the connection between Bosnia and the Muslim world have been noted. In spring 1999, clashes between Serbs and ethnic Albanians in the Serbian territory of Kosovo had led NATO to initiate intensive air strikes against Serbia in an attempt to force the Belgrade authorities to agree to the deployment of a NATO-led international peacekeeping force in Kosovo. Within NATO itself, there are serious tensions between Greece and Turkey over Cyprus and other issues in the Aegean. It is not the intention here to examine the various disputes between the two NATO allies. Suffice it to say that a conflict between Greece and Turkey would have serious consequences for NATO's role in the Mediterranean.

With regard to relations between states north and south of the Mediterranean, there is a degree of apprehension among NATO states along the periphery of Europe concerning the possible threat posed by the proliferation of ballistic missiles and the spread of weapons of mass destruction (WMD) in the south. Turkey continues to monitor closely any upgrading of the Syrian armed forces and also keeps a watchful eye on Iran and Iraq. In North Africa, Libya under the erratic leadership of Mu'ammar Gadhafi remains a potential source of danger. In 1986, after a US air strike against Libya, Gadhafi retaliated by attempting to hit the Italian island of Lampedusa with SCUD missiles.

NATO states bordering or near the Mediterranean are also concerned about other risks and challenges that stem from the south. The possible impact of the spread of radical, politicized Islam is a cause for alarm. France and Turkey have been the victims of acts of terrorism perpetrated by Muslim extremists. French officials, in particular, are fearful that the civil unrest in Algeria may spill over and affect the large Muslim immigrant community in France. Governments in France, Italy, Portugal, and Spain are anxious to contain the increasing number of migrants from North Africa who are hoping to escape violence and political repression at home and seek employment and a better standard of living in the north. Nationalist politicians in France and Italy, in particular, warn that the continued arrival of Muslims from the southern Mediterranean threatens to destroy the social fabric of society in southern Europe. The economic, social, environmental, and demographic problems of what are largely impoverished societies in North Africa ruled by authoritarian governments could pose serious problems for the security and stability of the Mediterranean as a whole.

Some officials in NATO have referred to a "threat from the south." This, in turn, has influenced how Arab governments and their peoples view the north. The Arab public, in general, is very suspicious and critical of Europe and the United States. Most Arab governments, more aware of the need to maintain close economic and political ties with the EU and the United States, at the same time have to take into account the feelings and concerns of their peoples. Many in the Arab world regard Europe and the United States as anti-Arab, anti-Muslim, and pro-Israel. Ideological differences, cultural distinctions, and past history—the Crusades, and more recently Western colonialism—help in part to explain the Arab mistrust of Europe and the United States. The West (or north as it were) is accused of double standards. Arabs believe that it is unjust that tough measures are taken against Iraq and Libya while no United Nations (UN) sanctions are imposed on Israel for violating UN Security Council resolutions. Many Arabs suspect that the West is seeking to interfere in the internal affairs of Arab states and impose alien values on their societies. For many in the south, there is a fear of a threat from the north. However, Arab governments—excepting Libya and Sudan (after the US missile strike against a suspected chemical weapons factory near Khartoum in 1998), and excluding the separate case of Iraq—are aware that Europe and the United States are not about to launch an attack against an Arab state. Nevertheless, while popular suspicion and mistrust of the West prevails, Arab officials cannot be seen to be cooperating too closely with their Western counterparts.

In reality, conflict is more likely to break out between states in the south. Ballistic missiles and possibly WMD would more probably be deployed in a south-south than north-south military encounter. An Arab-Israeli conflict is still possible in spite of the beginning of a Middle East peace process. This peace process is discussed separately in the next section. Possible inter-Arab conflict must also be considered. As of early 1999, the problem of Western Sahara was still not resolved. A renewal of the conflict between the Moroccan armed forces and the Algerian-backed forces of the Popular Front for the Liberation of Saguia el-Hamra and Rio de Oro (POLISARIO) is possible if a permanent solution for the Western Sahara that is satisfactory to all parties is not found. Morocco and Mauritania disagree over the ownership of certain territories. Egypt suspects that Sudan is providing training for radical Egyptian Islamic groups. In the years to come Libya may again create problems for its neighbors. A number of border disputes between North African states have been resolved in the 1970s and 1980s. However, the failure to demarcate certain maritime boundaries, fishing areas, and economic zones could lead in future to serious disputes if not conflict.

Interstate disputes in the south may have an impact on north-south relations in the Mediterranean area. In the Gulf region, it was the Iraqi invasion of Kuwait in 1990 that triggered the direct involvement of many NATO member states in Operation Desert Storm. Potential interstate disputes in the south cannot be disregarded given NATO's interest in the oil and gas reserves of North Africa and the Middle East. The importance of the Middle East peace process, and the Alliance's concern for the safety of Western civilians living and working in the southern Mediterranean, often in the oil and gas industries, must also be considered.

Intrastate problems in particular states in the south may also have an impact on the stability and security of the Mediterranean area. Deterioration in relations between Israelis and Palestinians in Israel could have widespread destabilizing consequences. Ruling elites in the Arab world are determined to remain in power. The legitimacy of their rule may be increasingly challenged by publics less willing to tolerate economic and social inequalities. Arab leaders, mindful of the example of Algeria, are fearful of the spread of radical Islam. There are demands in some quarters for popular participation in decision-making and for politicians to be held more accountable for their actions. Ruling Arab elites are realizing that they must listen to the opinions of their publics. The difficulties this may pose for relations between states north and south of the Mediterranean has been previously noted.

There are a number of problems and potential sources of dispute in the Mediterranean that must be addressed. Preventive diplomacy, confidence building, and dialogue may contribute toward making the Mediterranean a more stable and secure area. Although NATO's Mediterranean Initiative focuses on relations between states north and south of the Mediterranean, it may also indirectly encourage improved relations between states in the southern and eastern Mediterranean, and could even have a positive impact on internal developments in particular Arab states. In the Mediterranean, the EU's Barcelona Process is more directly interested in interstate and intrastate problems in the south, as well as with north-south relations.[18] Both NATO and the EU, however, consciously steer clear of the Arab-Israeli problem in order not to create complications for the Middle East peace process. The Middle East peace process is a huge subject that can not be extensively discussed in this study. But the future of NATO's Mediterranean Initiative and the EU's Barcelona Process are largely dependent on the success or failure of this peace process. It is thus important to have some understanding of the peace process. The next section briefly examines the problems that the peace process has encountered since the formation of the Likud-led coalition government in Israel in May 1996.

THE MIDDLE EAST PEACE PROCESS

Until the launching of the Middle East peace process in the early 1990s, most Arab governments and peoples had refused to acknowledge the existence of the state of Israel. With the end of the Cold War, however, radical Arab states and movements were no longer supported by the Soviet Union. Syria even participated in the US-led international coalition to oust occupying Iraqi forces from Kuwait. In these new circumstances, and with the support of the international community in general, the United States and the Soviet Union cosponsored a meeting in Madrid in October 1991 of all the major parties in the Middle East. This led to the start of direct bilateral negotiations between Israel and the Palestinians, Israel and Jordan, Israel and Lebanon, and Israel and Syria. Also, multilateral talks commenced in the form of working groups that aimed to address various issues and problems in the Middle East. Progress in talks between the Israelis and Palestinians was achieved in Oslo. In September 1993 in Washington, D.C., the Israeli prime minister, Yitzhak Rabin, and the chairman of the Palestine Liberation Organization (PLO), Yasser Arafat, signed a Declaration of Principles, which aimed for the gradual normalization of relations between the Israeli government and the Palestinians. Palestinian autonomy in parts of the West Bank and the Gaza Strip was agreed upon. A Palestinian civil administration was to be formed. A second agreement, signed in Washington, D.C., in September 1995, envisioned the Palestinians gradually taking control of other territory in the West Bank as Israeli troops withdrew or redeployed. Earlier, in October 1994, Israel and Jordan had concluded a peace treaty.

In spite of these positive developments, relations between Israel and the Arab world were still far from smooth. Israel and Syria were unable to reach agreement on the future of the Golan Heights. There was little progress in the talks between Israel and Lebanon. Israeli forces were still occupying southern Lebanon in order to ward off attacks from radical pro-Iranian Hezbollah guerrillas. With Egypt at the forefront, Arab states were determined that Israel should dismantle its nuclear arsenal. Arab governments argued that they had a right to develop WMD as long as Israel possessed nuclear weapons.

The peace process began to encounter serious difficulties after the assassination of Prime Minister Rabin by a Jewish extremist in early November 1995. There was an upsurge of violence within Israel in February and March 1996 orchestrated by Hamas, a militant Palestinian group that was critical of the peace process. In spring 1996, in the Grapes of Wrath Operation, Israeli air forces retaliated against Hezbollah attacks in southern Lebanon by launching an aerial bombardment that resulted in the deaths of over one hundred Lebanese civilians sheltering in a UN camp at Cana. Feeling increasingly under siege, the Israeli electorate voted in a Likud-led coalition government in May 1996. The new prime minister, Binyamin Netanyahu, was dependent on support from religious and ultraorthodox parties who wanted to adopt a tougher stance against the Palestinians. The peace process almost came to a standstill. Netanyahu managed to hold his first meeting with Arafat only in September 1996.

Escalating violence in Israel resulted in delays in the promised Israeli troop withdrawals from parts of the West Bank. In September 1996 the Netanyahu government decided to reopen a tunnel in the old city of Jerusalem near the Al-Aqsa mosque. Palestinians rioted and Israeli and Palestinian security forces clashed. The UN Security Council denounced the decision of the Israeli government. Tensions increased after the Netanyahu government decided in February 1997 to begin the construction of a new Jewish settlement at Har Homa (known as Jabal Abu Ghunaym to the Palestinains) in east Jerusalem. The Palestinians believed that the Israelis were undermining the spirit of the earlier Oslo accords. It appeared that the Israelis were determined to increase their presence in east Jerusalem. The UN General Assembly condemned the Israeli decision. Negotiations between the Israelis and Palestinians were suspended for eight months. In the meantime, in July 1997, thirteen people were killed in the busy Jewish market in Jerusalem as the result of the action of a Palestinian suicide bomber.

The peace process did not completely break down. The United States and the EU, and the international community in general, fearful of the potentially destabilizing consequences for the Middle East, were anxious to ensure that violence between Israelis and Palestinians did not spiral out of control. Negotiations would thus resume. Hopes were pinned on building on the success of the Hebron Agreement of January 1997. Israelis and Palestinians had been able to come to a compromise concerning a continued but reduced Israeli troop presence in Hebron there to protect the small number of Jewish settlers who refused to leave the town.

The peace process briefly acquired new momentum when in October 1998, with the mediation of the Clinton administration, Netanyahu and Arafat signed the Wye River Accord in the United States. With this "land-for-security" agreement, a timetable was fixed for two further redeployments of Israeli forces in the West Bank. Israel promised to withdraw from 13 percent of the West Bank. That would result in 40 percent of the West Bank coming under either the sole jurisdiction of the Palestinian Authority or joint jurisdiction with Israel. Netanyahu also promised to release several hundred Palestinian "political prisoners," allow the opening of a Palestinian airport in Gaza, and guarantee the safe passage for Palestinians moving between Gaza and other Palestinian areas. In return, the Palestinians again agreed to annul the clauses of the PLO Founding Charter that called for the destruction of Israel. They also promised to confiscate illegal weapons and take firm measures against terrorist organizations such as Hamas. Arafat also pledged not to declare unilaterally an independent Palestinian state on May 4, 1999. This date marked the end of the five-year interim period agreed to at Oslo, by which time a permanent agreement should have been concluded.

By December 1998 the implementation of the Wye Accord had been suspended after Israel had withdrawn only from a further 2 percent of the West Bank. The airport at Gaza, though, had been officially opened. The Netanyahu government complained that the Palestinians were not fulfilling their obligations concerning the confiscation of weaponry and the clamping down on the incitement of violence against Israel. The Palestinians were furious that the Israeli authorities had released mostly common criminals rather than political prisoners. At the time of writing, the Netanyahu government had lost a vote of no confidence in the Knesset and elections were scheduled for May 1999. Arafat was considering declaring an independent Palestinian state on May 4. Israeli officials were threatening to retaliate by annexing parts of the West Bank. Other issues remained unresolved. Safe passage between the Gaza Strip and the West Bank was still not guaranteed. No progress had been made with regard to what to do with four million displaced Palestinian refugees. The future and final status of Jerusalem—an issue of keen interest to the Muslim world in general—remained to be tackled.

The initial positive developments in Israeli-Palestinian relations had resulted in Arab states effectively ending their economic boycott of Israel. Several Arab governments had decided to establish low-level diplomatic ties with Israel. However, the decision of the Netanyahu government to go ahead with the building of the Har Homa settlement led to deteriorating Arab-Israeli relations. At the end of March 1997 the Arab League called for a freezing of the normalization of relations with Israel. The Egyptians, Jordanians, and Palestinians were not directly affected because of earlier treaties they had concluded with Israel. However, the Egyptians and Palestinians refused to send delegations to the fourth Middle East–North Africa (MENA) economic conference held at Doha in Qatar in November 1997. Only eight Arab states—including Jordan—decided to attend, because of the presence of the Israelis at Doha. Most Arab states in the southern and eastern Mediterranean and the Gulf region, together with Israel, had participated in previous MENA conferences. At the time of writing, there was no progress in relations between Israel and Syria and Lebanon, in spite of the decision of the Netanyahu government in spring 1998 to accept UN Security Council Resolution 425 with regard to the withdrawal of Israeli forces from southern Lebanon. The Syrians and Lebanese objected to what they believed were unnecessary Israeli preconditions before the withdrawal could take place. Many Arabs were accusing the United States of double standards. The Clinton administration had bombed Iraq in December 1998 after the work of UN weapons-inspection teams had been impeded. But, according to most Arabs, the United States was doing little to pressure Israel to conform to UN Security Council resolutions regarding the occupied territories.

The original hopes and expectations generated by the launching of the Middle East peace process had been dashed although not quite shattered by early 1999. The United States and the EU, in particular, were still encouraging the Israelis and Palestinians to settle their differences. Israeli-Jordanian relations had weathered a crisis in September 1997 when Israeli agents had botched an attempt to assassinate a Hamas leader in Jordan. It was hoped that the suspension of the activities of the multilateral working groups, set up after the Madrid Summit to deal with the various issues in the Middle East, would be only temporary. However, the situation within Israel remained extremely fragile. Given these circumstances, there was a feeling in the United States and in Europe that initiatives seeking to improve security in the Mediterranean as a whole—such as the Barcelona Process and the NATO-Mediterranean dialogue—should carefully avoid becoming directly involved in the Middle East peace process. There was a general concern that these initiatives should not be seen as interfering with—and, perhaps, thereby complicating—the peace process. In spite of the support of the international community, there was a danger that the Middle East peace process could completely collapse.

THE NEED FOR PREVENTIVE DIPLOMACY

"Preventive diplomacy," as defined by former UN Secretary General Boutros Boutros-Ghali, consists of "action to prevent disputes from arising between parties, to prevent existing disputes from escalating into conflicts and to limit the spread of the latter when they occur." Boutros-Ghali noted that the UN and regional organizations could be engaged in preventive diplomacy. These organizations could implement various confidence-building measures. These might entail the exchange of military missions and the formation of regional/subregional risk-reduction centers; fact-finding missions; early warning systems to pick up information about a possible conflict; the preventive deployment of, for example, UN troops to discourage hostilities; and the establishment of demilitarized zones.[19]

Michael S. Lund has criticized this definition of preventive diplomacy on the grounds that it is too inclusive. Three levels of a conflict or potential conflict are identified by Boutros-Ghali. The first part of Boutros-Ghali's definition, according to Lund, refers to dispute prevention. At this level or stage it is necessary to tackle issues such as poverty, environmental degradation, the inadequate provision of education, nationalism, which would require "generalized policies" like plans for income redistribution, family planning, universal education, and disarmament programs. For Lund, these policies are too broad to warrant the label preventive diplomacy. Lund further argues that the final part of Boutros-Ghali's definition of preventive diplomacy is actually concerned with violence containment and is thus part of crisis management and policies aimed at terminating a war. According to Lund, one should therefore concentrate on the middle element of Boutros-Ghali's definition. A more accurate definition of preventive diplomacy should focus on how to prevent low-intensity disputes from becoming high-intensity confrontations possibly involving the use of armed forces.

Thus, preventive diplomacy operates in between measures for long-term societal and global betterment on one hand, and short-term crisis management on the other.[20]

According to Lund, a more promising definition of preventive diplomacy for the post–Cold War is as follows:

> Action taken in vulnerable places and times to avoid the threat or use of armed force and related forms of coercion by states or groups to settle the political disputes that can arise from the destabilizing effects of economic, social, political, and international change.[21]

In Lund's opinion, governments, multilateral organizations, NGOs, individuals, and the disputing parties themselves may be involved in preventive diplomacy by making use of diplomatic, political, military, economic, and other instruments. In those situations where the socio-economic conditions are serious enough to be a cause of violence, economic instruments should be employed. Lund refers to how important it is to address issues in dispute by engaging the parties in dialogue or negotiation. This would help to modify "perceptions and feelings of mistrust and suspicion among the parties."[22]

Lund argues that preventive diplomacy would be required "when tensions between parties are in danger of shifting from stable peace to unstable peace or worse." He then states that "*preventive diplomacy is especially operative at the level of unstable peace*" [original in italics]. In a "stable" or "cold" peace there is limited cooperation and communication between parties within an overall basic order. There are differences in values and goals among the parties, but disputes are usually resolved by non-violent and often predictable means. In a situation of "unstable" or "negative" peace, there is a high degree of tension and suspicion among the parties and there may even be sporadic acts of violence. The parties perceive one another as enemies and maintain armed forces to deter a possible attack.[23]

Using these criteria, how may one evaluate the security situation in the Mediterranean in general in the late 1990s? Concerning interstate relations in the north, crisis diplomacy and crisis management still seemed to be the order of the day in the Balkans. Relations between the NATO allies, Turkey and Greece, were bordering between stable and unstable peace. With regard to the interaction between states north and south of the Mediterranean, in general a stable or cold peace prevailed. Likewise, on the whole, in relations between Arab states there was a situation of stable or cold peace. In many instances, though, Arab-Israeli relations were on the verge of deteriorating into a situation of unstable or negative peace. If the Middle East peace process totally collapsed there would be a renewed crisis in the area. Concerning intrastate relations in the southern and eastern Mediterranean, there was civil war in Algeria. In many other Arab states political, social, and economic conditions may continue to worsen. Within Israel and the occupied territories, tensions between the Palestinians and Israelis threatened to lead to a serious crisis. These intrastate problems could have a negative impact on the general security situation in the Mediterranean.

The NATO-Mediterranean dialogue, with its focus on the relations between states north and south of the Mediterranean, is thus involved with a situation in which a stable or cold peace prevails. It will be argued that it is much more difficult to commence a dialogue between parties in a situation of unstable or negative peace when the parties are more likely to perceive one another as enemies. One may contend that the NATO-Mediterranean dialogue, in effect, is aiming to ensure that the relationship between states north and south of the Mediterranean does not become one in which an unstable or negative peace dominates. One may add that NATO officials are seeking to promote the development of a "warm" and durable peace between states in the Mediterranean area. According to Lund, in a situation where so-called warm peace prevails, the governments of states believe in the same values and share common goals. Disputes are settled peacefully through institutionalized channels. A high level of cooperation and economic interdependence encourages the sense of international "community."[24] Ideally, NATO officials would also hope that their Mediterranean Initiative could help promote the development of a warm peace between Arab states and between Arab states and Israel, and also improve "security" and "stability" within certain states in the Mediterranean area.

Therefore, with regard to relations between states north and south of the Mediterranean in particular, preventive diplomacy may be needed to maintain a stable peace when there is a real possibility that without such diplomacy a situation of unstable peace instead could emerge. Preventive diplomacy is required in such circumstances, although, as previously noted, Lund stated that preventive diplomacy is mostly applicable at a time of unstable peace.

CONCLUSION

The Mediterranean is an area in which there are a number of disputes that could erupt into conflict if the issues are not successfully addressed.

And certain disputes could break out in renewed conflict—the Western Sahara and Arab-Israeli cases—if they are not properly resolved. The NATO-Mediterranean dialogue is seeking to improve relations between states north and south of the Mediterranean. It is a form of preventive diplomacy. The future success of this dialogue is to a large extent dependent on more stable relations developing between southern Mediterranean states, and especially between Arab states and Israel. The Middle East peace process, in turn, is dependent on Israelis and Palestinians resolving their differences and working out a final and mutually satisfactory peace settlement.

Significantly, the NATO-Mediterranean dialogue does not include states that may pose more of a real security threat to NATO member states. Libya and Algeria are excluded. Governments in southern Europe would likely feel uneasy if a radical, Islamic government were established in Algiers. It will be observed later in this study that some Arab officials have questioned the effectiveness of NATO's Mediterranean Initiative because of its exclusion of certain states.

In order to assess the possible value of the NATO-Mediterranean dialogue it is important to examine what is precisely meant by "security" and "stability" when referring to the Mediterranean. There is a debate over what is meant by the term "security." Security appears to be perceived differently by governments and peoples north and south of the Mediterranean. Cultural differences to some extent affect how groups perceive one another and may thus have an impact on security issues. The Arab public, in general, appears to regard with much suspicion the United States and western Europe. It will be demonstrated that "security" and "stability" are not one and the same even though the terms are often employed interchangeably. Continued stability may be at odds with enhanced security. The next chapter will include a discussion of the role of confidence-building measures (CBMs) and so-called confidence- and security-building measures (CSBMs) in the Mediterranean area. These are important elements of preventive diplomacy. A dialogue may be a useful CBM. The NATO-Mediterranean dialogue will be analyzed from this perspective. However, while the NATO-Mediterranean dialogue focuses on north-south interstate relations, Arab governments are more concerned with south-south interstate relations and with "security" and "stability" problems within their state. These concerns of Arab leaders would seem to set limits and place bounds on the possible effectiveness of the NATO-Mediterranean dialogue.

NOTES

[1]John Chipman, "Introduction," in *NATO's Southern Allies: Internal and External Challenges,* John Chipman, ed. (New York and London: Routledge, 1988), 3.

[2]Roberto Aliboni, *European Security across the Mediterranean* (Paris: Chaillot Papers 2, Institute for Security Studies, WEU, 1991), 2–3.

[3]Ian O. Lesser, "Southern Europe and the Maghreb: US Interests and Policy Perspectives" (paper elaborating on remarks delivered at the conference Employment, Economic Development and Migration: European and US Perspectives, organized by the Luso-American Development Foundation, Lisbon, April 28, 1995), 7–8.

[4]Louis J. Cantori and Steven L. Spiegel, *The International Politics of Regions: A Comparative Approach* (Englewood Cliffs, N.J.: Prentice-Hall, 1970), 6–7.

[5]Ibid.

[6]W. Howard Wriggins, ed., *Dynamics of Regional Politics: Four Systems of the Indian Ocean Rim* (New York: Columbia University Press, 1992), 6–8.

[7]Barry Buzan, *People, States and Fear: The National Security Problem in International Relations* (Chapel Hill: University of North Carolina Press, 1983), 106.

[8]Andrew Hurrell, "Latin America in the New World Order: A Regional Bloc of the Americas," *International Affairs* 48, 1 (January 1992): 123.

[9]Iver B. Neumann, "A Region-Building Approach to Northern Europe," *Review of International Studies* 20, 1 (January 1994): 58–59.

[10]For an argument along these lines, see Ole Waever, "Nordic Nostalgia: Northern Europe after the Cold War," *International Affairs* 68, 1 (January 1992): 99–100.

[11]Neumann, "A Region-Building Approach to Northern Europe," 66, 69.

[12]Thomas P. Thornton, *The Challenge to US Policy in the Third World: Global Responsibilities and Regional Devolution* (Boulder, Colo.: Westview, 1986), 25.

[13]Fernand Braudel, *The Mediterranean and the Mediterranean World in the Age of Philip II* (New York: Harper and Row, 1972).

[14]Aliboni, *European Security across the Mediterranean,* 1–2.

[15]Stephen C. Calleya, *Navigating Regional Dynamics in the Post–Cold War World: Patterns of Relations in the Mediterranean Area* (Brookfield, Vt.: Aldershot: Dartmouth, 1997), 235.

[16]Stephen C. Calleya, "Post–Cold War Regional Dynamics in the Mediterranean Area," *Mediterranean Quarterly* 7, 3 (summer 1996): 44.

[17] Fred Tanner, "The Mediterranean Pact: A Framework for Soft Security Cooperation," *Perceptions* 1, 4 (December 1996–February 1997): 56.

[18] For an alternative viewpoint that argues that the Barcelona Process does not pay enough attention to interstate and intrastate problems in the southern and eastern Mediterranean, see Roberto Aliboni, "Confidence-Building, Conflict Prevention, and Arms Control in the Euro-Mediterranean Partnership," *Perceptions* 2, 4 (December 1997–February 1998): 73–86.

[19] Boutros Boutros-Ghali, *An Agenda for Peace—Preventive Diplomacy, Peacemaking and Peace-Keeping—Report of the Secretary-General Pursuant to the Statement Adopted by the Summit Meeting of the Security Council on 31 January 1992* (New York: United Nations, 1992), 11, 13–19.

[20] Michael S. Lund, *Preventing Violent Conflicts—A Strategy for Preventive Diplomacy* (Washington, D.C.: US Institute of Peace, 1996), 34–37.

[21] Ibid., 37.
[22] Ibid., 37, 45.
[23] Ibid., 40–41, 39.
[24] Ibid., 39.

CHAPTER 2

The Importance of "Dialogue" as a "Confidence-Building Measure"

INTRODUCTION

NATO's Mediterranean Initiative launched what was in effect a dialogue focusing on security issues between the governments of states north and south of the Mediterranean. Soon, NATO officials became interested in incorporating other nongovernmental groups into their Mediterranean dialogue. In order to have a better understanding of how effective this initiative may be, one must first be clear about what a dialogue is and what are the preconditions required before a dialogue might commence. Dialogue is an important part of preventive diplomacy, as already noted. It is also an invaluable CBM.

This chapter will consist first of a discussion, though, of what is meant by "security," and also of an examination of how "security" and "stability" interrelate in the context of the Mediterranean. What does one mean exactly by Mediterranean security? In the previous chapter it was noted that one may refer to the security of a region or area, or one may focus on relations between states in the area or between groups and individuals within states in the area. Whose security is one thus aiming to discuss? Another complicating factor is that security issues are perceived differently by individuals and groups north and south of the Mediterranean. Differences in culture may also come into play here. These questions need to be raised and discussed because they have significant repercussions on the relationship between "security" and "stability" in the Mediterranean. They also have a direct bearing on the possible impact of NATO's Mediterranean Initiative.

WHAT IS MEANT BY "SECURITY"?

There has been much discussion about what is meant by "security." According to one commentator, it is a "notoriously amorphous concept."[1] Security is also regarded as a "contested concept." What is it that is to be made secure?[2] Security is vague in its content and format. One scholar has questioned: "Is it a goal, an issue-area, a concept, a research program, or a discipline?"[3] This study examines security as a concept and as a goal. The aim of the NATO-Mediterranean dialogue is to enhance security—however security may be defined—in the Mediterranean area.

Traditionally, analysts assumed that states were the objects to be made secure. Studies on security focused on military and defense issues such as arms control, terrorism, the proliferation of WMD. These are often referred to now as hard security issues. Value was placed on stability and on maintaining the status quo.[4] Within the framework of the so-called security dilemma, enhancing the security of one state by expanding its armed forces could be perceived as threatening by other states. Security was thus viewed as "essentially competitive." It was concerned with countering possible "external" threats.[5]

Some commentators in recent years argue that security does not always need to be competitive. It is becoming more commonplace for analysts to refer to the possibility of rather promoting "shared," "common," or "cooperative" security between states. In the post–Cold War era the world has become more interdependent. Institutions and regimes—the term "regime," in this context, will be discussed later—may now have a more decisive and positive influence on global politics.[6] However, as previously noted, the Mediterranean area is one in which there are tensions among and between states both north and south of the Mediterranean. In the 1990s, therefore, the notion of cooperative security is still of little relevance when focusing on interstate relations in the Mediterranean. It is much more appropriate, for example, when examining relations between states in western Europe.

Advocates of new thinking on security in the post–Cold War era also emphasize that the focus of attention should no longer be on only external, military threats to states. They argue, rather, that there is now a need to also include potential threats stemming from other areas—economic, environmental, societal—that are often referred to as areas of soft security. According to Buzan, security has become a "multidimensional concept." Security operates in several dimensions—military, political, economic, societal, and environmental—that may interrelate and interact. Political security "concerns the organisational stability of states, systems of government, and the ideologies that give them legitimacy." Economic security "concerns access to the resources, finance and markets necessary to sustain acceptable levels of welfare and state power." Societal security "concerns the sustainability, within acceptable conditions, for evolution of traditional patterns of language, culture and both religion and national identity and custom." Environmental security "concerns the maintenance of the planetary biosphere as the essential support system on which all other human enterprises depend."[7]

Security is concerned with threats and vulnerabilities. Perceptions are here a crucial but also complicating factor. Buzan has been criticized for overextending the concept of security. It has been suggested that the emphasis should only be on supposed vital threats.[8] Of course, nonvital threats can develop to become vital threats. And individuals or governments may perceive threats as vital when in practice they are not. In this context, the distinction between issues of hard and soft security may become less obvious. There may be military, economic, environmental, societal threats, and so forth, that could be vital and thus be a threat to the security of a society or community.

With regard to the Mediterranean, one may argue that concerning interstate relations, soft and hard security issues are important. New thinking on security, however, also underlines that the security of individuals and groups, and regional and global security, are as much a matter of concern as the security of states. The issue of regional security was raised in the previous chapter. Concerning the Mediterranean, the security of individuals and the relations between a government and its citizens need to be considered. Here, the issues of human rights, domestic stability, and political legitimacy come to the fore.[9]

In Buzan's opinion, issues of societal security concerning the threats and vulnerabilities affecting patterns of communal identity and culture are becoming increasingly significant.[10] The security of certain groups, as well as of states and individuals should therefore be considered. In the context of societal security, Buzan refers to the impact of migration and the possible clash of rival civilizational identities, most notably between the West and Islam. These issues will be discussed later.

In well-established, mature democracies, attention on security is less likely to be concentrated on societal security. Individuals and groups would most probably be able to be peacefully accommodated. Threats would stem from an external source. These threats could be military in nature, but they might also be societal—perhaps from a flood of migrants—economic, or environmental. In less-developed states controlled by authoritarian governments, threats are more likely to stem from internal as well as external sources. These governments might lack legitimacy at home. They could be crippled by substantial public debts and might be the victims of an ecological catastrophe. In developing states many conflicts or disputes are intrastate rather than interstate, as unpopular elites are often struggling to cling to power.[11] Threats to these elites are usually political, societal, and/or economic in nature.

The security concerns of elites in the developing world who are determined to maintain a hold on power has been explored by Mohammed Ayoob. He focuses on how these elites feel vulnerable and perceive threats. These elites rule postcolonial states in which the process of state building is far from complete. In some instances, therefore, they are fearful that minority ethnic groups might attempt to play the cards of human rights and self-determination in an effort to secede. These elites are also concerned because their power bases are increasingly challenged by the demands from various groups for political and civil rights and greater economic benefits. In the post–Cold War world, ruling elites in developing states can no longer so readily call upon the major states to help them crush domestic opposition. Western governments in particular are pushing for democratization and the improvement of human rights in the developing world. This places additional pressures on the ruling elites of less-developed states. Examining these issues from the perspective of these ruling elites, Ayoob notes that "internal insecurities fundamentally determine the security predicament of the Third World states."[12]

Ayoob recommends that ruling elites in the developing states sincerely seek to accommodate the interests of minority groups, guarantee basic human rights for individuals, and assure the political participation of all citizens. In Ayoob's opinion, these actions would also help to consolidate the state in question.[13] In practice, though, governing groups may strive to preserve their monopoly on power by endeavoring to bolster their legitimacy through other means. For example, a ruling elite may seek to strengthen its position by manufacturing a crisis with another state in order to distract its public from domestic problems. Or, in the case of the Mediterranean, for instance, the authorities could attempt to apportion blame for internal problems on the outside world and on the West in particular. This ploy could rebound against those ruling elites who are dependent to some extent on Western economic and political support. Stoking the flames of anti-Western sentiment could also whip up political or religious radicalism in the developing state, thus endangering the position of the current ruling elites.

Elites in the developing Arab world are in most instances, therefore, primarily concerned with threats to intrastate security. Security issues concerning inter-Arab or Arab-Israeli relations are of secondary importance. The north (or West) is usually not regarded as a serious security threat by ruling Arab elites. There is a cultural divide between those north and south of the Mediterranean and this, as will be seen, does have an impact on security perceptions of certain groups. However, the importance of cultural distinctions should not be overplayed. Ruling Arab elites may choose to make use of cultural differences between societies north and south of the Mediterranean in order to win popular support and bolster their positions. However, as noted, the governing authorities could be involved in a potentially dangerous exercise. These elites would usually seek neither to alienate the outside world nor to encourage the forces of political extremism at home.

In spite of many joint references in speeches and texts to the importance of "security" and "stability," in practice, security and stability may actually be at odds with each other. As noted, traditionally, security in the Cold War era was primarily concerned with maintaining stability and the status quo. In the post–Cold War era a totally new set of circumstances has emerged. Previously, the authoritarian leaders of the Third World were able to secure support and bolster their standing by cultivating ties with either NATO or Warsaw Pact states. These leaders are now under increasing pressure from outside (from the West in particular) and from below (from their own publics) to introduce serious political and economic reforms. There is a possibility of change. A generally accepted view is that more liberal and democratic governments are much less likely to pursue aggressive and expansionist policies. One problem, though, is that moves toward democratization in states that have previously been under firm authoritarian control could lead to a period of instability in the state or states in question and adversely affect the security of neighboring states. Instability in the short term, though, may enhance the security and stability of a state and surrounding states in the longer term. It is important to realize that the stability of an authoritarian government is not necessarily beneficial to the security of individuals and groups within that state and is not necessarily conducive to the maintenance of regional stability and security. The case of the Iraqi government under Saddam Hussein is illustrative.

With the end of the Cold War, policy-makers in NATO have undertaken a reassessment of what they mean by security. This has led to the acceptance of a broadened and enlarged concept of security. The launching of the NATO-Mediterranean dialogue should be understood within this context. One must take into account the military, political, economic, societal, and environmental dimensions of security in the Mediterranean area. NATO officials should be aware that these aspects of security are becoming more important for ruling elites in North Africa and the Middle East as these elites are under increasing pressure to introduce domestic reforms. NATO as an organization, though, is better equipped to deal with interstate than intrastate relations.

Notions of "shared," "common," or "cooperative" security with regard to relations between states in the Mediterranean area are of little relevance at the present time. This is, in part, because the governments of western Europe and of the southern and eastern Mediterranean are different in nature. In the short term, at least, it seems that in certain less-developed states authoritarian elites may attempt to hold on to power for their own selfish security interests and ignore at their peril the security concerns of other individuals and groups. The notion of "cooperative" security also presupposes that different governments and publics have the same perceptions of what is meant by security. Here, the possible impact of cultural perspectives, which may not always coincide with viewpoints commonly held in the West, should be considered.

THE RELEVANCE OF "CULTURE"

What is meant by "culture"? Many social anthropologists believe that the term is too vague and has been employed so often that it has become almost meaningless. However, in political science and international relations the term may still be of use, particularly if it is looked on as involving rules governing interactions between individuals within a community. Culture might be regarded as a set of socially created and learned norms, standards, rules, or collective mental programming that makes a meaningful existence possible for members of a community. Here, one might refer to a "national culture" that might be contrasted and compared with various other national cultures.[14] Culture can thus affect the behavior of people and influence how they view and compare themselves to other peoples.

The term "culture" in international relations has had a recent negative press with alarmist accounts of the supposed threat of "the green peril" and the Islamic menace. Its use has also become a bone of controversy following Samuel Huntington's by now famous or infamous work on the clash of civilizations.[15] According to Huntington, culture has become a crucial element in the functioning of contemporary international politics, and the underlying basis for state action. In his opinion, civilization is the broadest level of cultural identity. Huntington contends that in the post–Cold War world the main cause of conflict will be cultural between Western and non-Western civilizations and in particular between the West and Islam because the most important element of a civilization is its religion. These views generated a series of counterarguments. It was noted how states controlled civilizations. Commentators also believed that economic rather than cultural factors were more significant.[16]

Cultural differences are striking in other ways. A culture provides a particular "perception-shaping lens" through which events are viewed and understood.[17] This is important to take into account when dealing with peoples of a different culture. The linkage between perception, threats, and security has been discussed. Work has been done on exploring the role of culture in the decision-making processes of states and their ruling elites. Culture is regarded as a "cognitive filter." Actors observe issues and interpret decisions through the prism of distinct cultural perceptions. Culture may be an obstacle to international understanding. It may have a negative impact on negotiations.[18] According to one commentator, culture "impinges upon negotiations by conditioning one's perception of reality, blocking out information inconsistent with culturally based assumptions, projecting meaning to the other party's words and actions, and leading a negotiator to an inaccurate attribution of motive."[19] Clearly, the possible impact of culture on a dialogue between individuals and groups coming from contrasting cultural backgrounds should be borne in mind.

Emphasis here is on the perceptions of elites. This is again so with reference to the notion of "strategic culture," which aims to underline how national security is influenced by cultural tradition. In the words of Alastair Ian Johnston: "Different states have different predominant strategic preferences that are rooted in the early or formative experiences of the state, and are influenced to some degree by the philosophical, political, cultural and cognitive characteristics of the state and its elites."[20] Within a state, however, one should also take into account the perception-shaping lenses of individuals and groups other than the ruling elite. The perspectives of minority ethnic groups or the public at large, for example, should not be dismissed. Serious problems may ensue if a ruling elite loses complete touch with the people on the street or in the countryside. Robert D. Putnam's depiction of a two-level game in the context of international negotiation may be of relevance here.

According to Putnam, when officials secure an agreement with representatives from another state—at the international level or level one— the approval of this agreement would also have to be obtained from the public at the national level (level two). Bargaining thus takes place at the national and international levels. Negotiators are concerned that they should secure the so-called win-set—the set of all possible international agreements that would gain majority support among domestic constituents. In practice, ruling elites would normally have to give primacy to level two as their continued incumbency often depends on maintaining support from below. Governing officials, therefore, when concluding international agreements, should be sensitive to how their publics may react to deals struck with other states or with international organizations or institutions.[21] Arguably, only the most authoritarian and dictatorial governments may be able to ignore or override public opinion.

It has been suggested that the real causes of conflict in today's world are primarily due to socioeconomic rather than civilizational factors. There is a reaction in some societies to the spread of Western democracy and consumerism. The rise of radical Islam is portrayed as one reaction. According to this line of argument, culture is viewed as a secondary force: "Culture becomes a tactic, a tool, not a fundamental cause of conflict itself." What is needed is to address socioeconomic circumstances rather than culture per se.[22] One should be careful, though, not to underestimate the importance of culture as a cognitive filter. To refer to culture as a mere tactic would appear to downplay the significance of cultural differences in today's world.

Concerning the security and stability of the Mediterranean area, one should neither overexaggerate nor underestimate the importance of cultural differences north and south of the Mediterranean. In reality, one may argue that the north is not seeking to impose a cultural hegemony over the south. The north is not aiming for the cultural homogenization of the Mediterranean. It will be seen that ruling Arab elites at times do perceive that the north (or West) is seeking to impose alien values on Arab societies. However, these same elites, aware of the value of economic and political support from the north, are still willing to cooperate with western Europe and the United States. Concentrating more on intrastate than interstate security concerns, Arab governing elites need outside backing to help bolster their legitimacy at home. One may contend, then, that the principal difference between how elites north and south of the Mediterranean perceive security threats is not based on cultural determinants. Compared to their rulers, however, in general, Arab publics are much more suspicious of the intentions of their northern neighbors and the United States. Here, cultural and historical factors come into play. It is important, therefore, that the Mediterranean dialogues launched by NATO, the EU, and the OSCE are also seeking to establish and develop contacts with peoples and not only with governmental officials. These dialogues may thus help to overcome and prevent further official and popular misperceptions that are at least in part—and more so in the case of Arab publics—due to cultural distinctions. Ideally, this should be a two-way dynamic with governments and publics on all shores of the Mediterranean reassessing their views of one another.

THE IMPORTANCE OF "DIALOGUE"

> *What does "dialogue" mean? The answer is that it means whatever you want it to mean. That is to say, it means nothing in particular. What politician would ever say that he was against it? And if no one can be against it, how can anyone be for it to any purpose?*[23]

This statement is somewhat cynical. Who would ever say that he or she was against peace? In reality, it is not always easy to commence a dialogue. Certain conditions must first be met. For example, two enemies are not likely to engage in a dialogue. One party may refuse to begin a dialogue with a party that may wish to initiate one. Politicians, in certain instances, declare that they are opposed to a dialogue.

In effect, a dialogue involves talks between two or more parties where the intention of those engaged in the dialogue should be to understand the positions of each party. This may then affect the future behavior of the parties toward one another. A monologue, on the other hand, takes place when one party monopolizes or dominates the proceedings. For a dialogue to be constructive, each party should be able to express its own positions and perspectives clearly and also listen attentively to the positions and perspectives of others—dialogue involves "open, effective communication."[24] The assumption, then, is that all the parties concerned have positive intentions and want to improve their relations with one another. The parties are neither enemies nor genuine allies. There are problems and difficulties in relations between the parties that they are seeking to address in the dialogue. Relations between genuine allies are much warmer. Genuine allies "consult" with one another or seek advice rather than engage in dialogue.

Therefore, a dialogue may commence when two or more parties are willing to talk to each other about their problems and difficulties in order to improve their relations through a better understanding and appreciation of the other party's (or parties') position and perspective. However, in practice, only one party may wish to enter a dialogue. Other parties may be opposed to a dialogue, perhaps believing that the party wanting to begin a dialogue is not sincere. Parties need to negotiate in good faith. No party should use a dialogue for propaganda purposes. A dialogue should also not be employed to gain intelligence information on another party. And, ideally, a dialogue should not be made use of by ruling authorities to divert the attention of public opinion from domestic difficulties.

Agreeing to enter into a dialogue entails risks. A party may feel it is becoming trapped when holding talks with other parties. These other parties may issue certain demands that may be impossible to meet.[25] There may be apprehension, then, that a party could attempt to impose its position through a dialogue. Or the party or parties objecting to a dialogue may believe that there is nothing to discuss. They may be of the opinion that their position on a certain issue or issues is correct and that the viewpoint of the other party or parties is wrong. Engaging in a dialogue in such circumstances could be seen as casting doubt on the validity of one's position. A dialogue might not commence because of disagreement over what specific issues should be discussed or due to a failure to reach consensus over what procedures should be adopted and/or at what level. One of the parties may prefer to take its case to a supposed impartial mediator who could, for example, be a representative of a state or an international body. In the case of a dialogue, therefore, ideally, all of the parties involved believe at the outset that they can benefit somehow from talking with one another. They are thus willing to engage voluntarily in a dialogue.

Open, effective communication, however, does not occur automatically. For communication to be effective, parties should be flexible and willing to change, modify, or adapt their positions and perspectives in the process of dialogue rather than rigidly hold to an original viewpoint or stance. Ideally, all parties should shift their original positions to some extent at least in order to demonstrate that one party is not dictating or dominating the dialogue. Opportunity should be provided for all parties engaged in the dialogue to express themselves openly. A dialogue is a form of negotiation. Negotiation often consists of making concessions, compromising, and bargaining. These elements may form part of the process of dialogue. It has been argued that negotiation, in its positive sense, is "a process of discovery which leads to some degree of reorganization and adjustments of understanding, expectations and behaviour, leading (if successful) eventually to more specific discussions about possible terms of a final, agreed outcome."[26] Thus, a dialogue may be regarded as a "process of discovery" between two or more parties. It may overcome misperceptions that may be historically rooted and/or the product of cultural factors.

Once a dialogue has commenced, in order for it to develop, all parties involved should contribute in some way to the dialogue. One party should not always assume the role of proposing initiatives while other parties adopt a purely passive role accepting or choosing not to accept these initiatives. This creates the impression that one party is monopolizing and dominating the dialogue, which is in danger of becoming a monologue. There is also a risk that in these circumstances the dialogue may come to a standstill if in the course of time the dominant party is unable to come up with fresh ideas about how the dialogue should proceed. Ideally, therefore, proposals and recommendations should be put forward by all parties. In the NATO-Mediterranean dialogue it will be seen that there is a tendency for NATO officials to lead the discussions while the representatives of the non-NATO Mediterranean countries appear to assume a more passive and reactive role.

Does a dialogue require definite goals and objectives? Or does it, rather, gradually evolve where the aims of the dialogue often shift in order to adjust to the changing positions of the parties involved? The dialogue will encounter difficulties if one or more parties complain that the dialogue should have clearly defined objectives. The ultimate goal of the NATO-Mediterranean dialogue is not spelled out. There are only general references to the need to improve stability and security in the Mediterranean. The non-NATO Mediterranean countries have complained about the lack of precise goals in NATO's Mediterranean Initiative. However, in practice, the NATO-Mediterranean dialogue has been able to evolve largely because of the success of the various cooperative activities closely associated with the dialogue. These activities lend to the dialogue a sense of movement and direction.

A dialogue should thus be a "process of discovery," but a "final agreed-upon outcome" is not guaranteed. Improved relations could suffice although this then prompts one to ask when a dialogue is adjudged successful. And when might a dialogue reach a "successful conclusion," bearing in mind that a dialogue cannot be expected to continue ad infinitum? The answers to these questions depend to an extent on the nature of the dialogue. Of course, a dialogue may collapse when one or more parties refuse to continue to negotiate, believing that there is no longer any benefit in prolonging talks. Or a dialogue may reach a successful conclusion when relations between the parties involved have improved through a better understanding of the perspectives of others and specific problems have been resolved. In these circumstances, through a gradual process, the dialogue may lose its initial prominence. As relations continue to "normalize," the original parties in the dialogue may start to relate to each other more as genuine allies, consulting and seeking the advice of others rather than aiming to tackle specific problems. In the course of time, therefore, a dialogue may imperceptibly and quite naturally evolve into another plane of communication. This process may occur because a dialogue, which is in effect a CBM, may encourage the development of other CBMs and also CSBMs, which could lead to the formation of a security regime, and ultimately, to the emergence of a so-called security community. These terms will be discussed shortly.

There are different types of dialogue. Dialogues vary according to the *nature* and *number* of parties participating in them. A dialogue involving two parties would obviously proceed along different lines from one consisting of twenty parties. Likewise, a dialogue would vary considerably from case to case if the participating parties were private individuals, businessmen, or state officials. Dialogues also differ according to the *type* and *range* of issues discussed. In terms of a "security dialogue," of which NATO's Mediterranean Initiative is an example, the parties may initially agree to address issues of soft security rather than tackle head-on hard security issues. A dialogue may also vary according to the *level* at which it is conducted. In the case of an interstate dialogue, talks may proceed, for example, between prominent or lower grade officials, or between parliamentarians or special experts attached to governments. Dialogues will also differ according to the *procedures* adopted. How often do the parties meet, where and for how long, and what specific procedures are followed when the parties gather? Discussions may be held on a bilateral or multilateral basis. Bilateral talks take place when only two parties are engaged in a dialogue. Multilateral talks occur when more than two parties participate simultaneously.

With regard to the number and nature of parties involved, the NATO-Mediterranean dialogue involves talks between, on the one hand, officials representing NATO, an organization that at the time of writing consisted of nineteen full-member states, and on the other hand, official representatives of six non-NATO Mediterranean states. It is concerned with enhancing the security and stability of the Mediterranean. As for procedural matters, and the level at which the dialogue has been conducted, the dialogue initially consisted of talks between officials of NATO's International Staff and representatives from the embassies in Brussels of the non-NATO Mediterranean states. NATO's International Staff received instructions from the North Atlantic Council (NAC). In turn, the members of the NAC acted upon the directives of their particular governments. NATO's International Staff held separate and bilateral rounds of discussion at NATO Headquarters in Brussels with representatives from the Brussels embassies of the non-NATO Mediterranean states. These meetings were voluntary, but it was intended that they should take place biannually. In practice, attendance at these meetings by non-NATO Mediterranean states has been irregular. These meetings were bilateral in the sense that NATO as an organization held talks on an individual basis with the non-NATO Mediterranean states. In 1997 NATO officials established the previously mentioned MCG to coordinate the NATO-Mediterranean dialogue. This raised the political level of the dialogue, as the MCG was composed of the political advisers of all of the NATO delegations based in Brussels. Once a year the MCG would meet with representatives of the non-NATO Mediterranean countries separately in a "16+1" format. The International Staff could also still meet with officials from the six dialogue countries.

In the course of the development of the dialogue, NATO officials began to hold discussions with so-called opinion leaders from the non-NATO Mediterranean countries. Meetings were thus organized, for example, with parliamentarians and academics. Contacts were thereby established with nongovernmental groups and individuals. Also, as part of the dialogue, various cooperative activities were drawn up between NATO officials and representatives from the six dialogue countries. Practical programs of cooperation were thus agreed on in the fields of information and science, and then also in the military sphere. These programs involved exchange visits, seminars, courses, and joint projects. NATO's Mediterranean Initiative thus dealt with issues related to the soft and harder aspects of security.

Once the NATO-Mediterranean dialogue became established, NATO's International Staff organized joint briefings to which representatives from the six non-NATO Mediterranean states were invited as a group. These briefings were held after sessions of the NAC, or were called to discuss specific issues of interest to the six non-NATO Mediterranean states. Was the NATO-Mediterranean dialogue gradually adopting certain multilateral features? The picture is a complicated one. There are major differences of opinion between Israel and the five Arab dialogue countries. There are also important divisions among the Arab countries. Furthermore, it should not be assumed that NATO itself functions smoothly as a united and monolithic organization. For example, there are disagreements between NATO member states bordering the Mediterranean and allies located farther to the north with regard to NATO's Mediterranean Initiative. The former grouping has generally taken a stance more in favor of developing closer links with North Africa and the Middle East.

CBMs AND CSBMs FOR THE MEDITERRANEAN

A CBM has been loosely and generally defined as "any step that decreases tensions or increases cooperation between states." CBMs are usually incremental in nature and emphasize the importance of openness or "transparency" in relations between states. They do not involve much risk taking. As a result of implementing CBMs, parties need not introduce major changes in their foreign policies or in the structure and size of their armed forces, for example. CBMs are thus "moderate as to their transaction costs."[27] CBMs usually address relations between states, although there is no reason why they cannot be employed for interactions between various groups and individuals within a state.

More specifically, CBMs have been defined by Marie-France Desjardins as actions falling into one or more of the following categories:

> exchanging information and/or increasing communication between the parties; exchanging observers and/or conducting inspections; establishing "rules of the road" for certain military operations; and applying restraints on the operation and readiness of military forces.[28]

The NATO-Mediterranean dialogue is certainly concerned with the first of these categories. If NATO's Mediterranean Initiative continues to build on what are basically military CBMs within its program of cooperative activities, the dialogue will also in the course of time incorporate elements of the other listed categories.

Desjardins has noted that in nearly all cases "CBMs are only as strong as the fundamental political will for compromise in any successful negotiations. Without preexisting detente CBMs are of little value."[29] It does appear that parties need to have a certain amount of confidence in one another before they begin to engage in confidence-building—building on this initial confidence and trust. This has been discussed with reference to the preconditions required in order for a dialogue to commence between parties. Dialogue may be regarded as one of the first steps in confidence-building. A dialogue is much more likely to begin when a situation of stable or cold peace rather than unstable or negative peace prevails between the parties concerned.

It has been suggested with regard to CSBMs, that certain "incentives" and "preconditions" are required before parties begin a negotiating process. The time needs to be "ripe" for this process to commence. This may be so when there is awareness among the parties of the costs of possible war and the risks of escalation of tensions. The time may also be "ripe," for example, when there is a realization that resources should be diverted toward dealing with economic, social, and ecological problems.[30] Benefits may then eventually accrue from embarking on a negotiating process. This argument may equally apply to CBMs as to CSBMs.

CBMs first received considerable international attention in 1975 when the Helsinki Final Act of the Conference on Security and Cooperation in Europe (CSCE) was released. According to the Preamble of the Document on Confidence-Building Measures of the Helsinki Final Act, CBMs should "contribute to reducing the dangers of armed conflict and of misunderstanding or miscalculation of military activities which could give rise to apprehension, particularly in a situation where the participating States lack clear and timely information."[31] The Act referred in particular, then, to military CBMs. Major military maneuvers conducted by either NATO or the Warsaw Pact had to be reported in advance. There would be prior notification of smaller military maneuvers and other military movements. Observers could be exchanged to attend maneuvers. Reciprocal exchanges of military personnel, including visits by military delegations between NATO and the Warsaw Pact, were to be promoted. The principal concern of the parties was to prevent a possible surprise attack by either of the military blocs. Initially, then, only very limited military CBMs were possible between NATO and the Warsaw Pact, as both blocs continued to perceive each other as enemies until the accession to power of Mikhail Gorbachev, after which relations dramatically improved. Originally, there was little or no mutual confidence and trust, and no progress was made on building on what were only preliminary military CBMs.

In time, CBMs may further develop and lead to the introduction of what have been referred to as CSBMs. This was the case in Europe with the 1986 Stockholm Conference on CSBMs and Disarmament in Europe and the follow-up Vienna documents of the Vienna Negotiations on CSBMs, produced in 1990, 1992, and 1994. The differences, if any, between CBMs and CSBMs are not clear. It has been argued that a key distinction between CBMs and CSBMs is that the latter are concerned with measures that are more "militarily significant, binding and verifiable." Others have provided a narrower definition of CSBMs, referring to them as "intentional and explicit cooperative measures that enhance openness and transparency in the security realm." According to this argument, CSBMs invariably deal with matters in the military domain. They must be directly negotiated and consensually agreed on by all parties.[32] In line with this definition, a dialogue cannot be a CSBM. In effect, though, CSBMs may actually be simply more advanced forms of military CBMs.

CSBMs, or more advanced military CBMs, are concerned with such matters as: the mandatory notification in advance of military exercises; the compulsory invitation of observers; and on-site inspection by challenge to verify compliance with regard to arms control measures concerning the size and nature of each party's armed forces.[33] Even more advanced CSBMs/CBMs deal with such matters as: the exchange of detailed information; the regular assessment of the implementation of CSBMs; the establishment of risk-reduction centers; and the creation of demilitarized zones or areas where military deployment is strictly limited. One may also include here mechanisms established to handle emergency situations. More openness, transparency, a greater exchange of information and verification are important elements of CSBMs or advanced military CBMs.

With the impact of Gorbachev's New Political Thinking and then the end of the Cold War, the CSCE and its successor, the OSCE, have been able to encourage the development of CBMs and CSBMs in Europe. This work is ongoing with, for example, the Forum for Security Cooperation of the OSCE playing an instrumental role.

This discussion has focused on what are referred to as the classic, traditional model, or the European model, of CBMs or CSBMs, with the emphasis on military and defense measures. This model was appropriate for two large military blocs opposing each other in Europe in the Cold War period. However, ruling elites in developing countries are pushing for the consideration of new types of CBMs that they believe will be more relevant to their needs. They are in favor of a more "comprehensive" approach to CBMs that will take into account threats from nontraditional security issues. These issues could include ethnic, religious, social, and economic disputes, for example.[34]

According to Desjardins, this input from the developing world has contributed to what is referred to as "new thinking" on CBMs. In this new thinking, there is more emphasis on confidence-building as a "process." How is confidence built on and strengthened? Dialogue, here, plays an important role. If parties debate, negotiate, and exchange information, this may help reduce mistrust. As the parties become better informed about each other, this may lead to positive changes concerning how each party perceives others. This could lead to a "spillover" effect prompting a general improvement in relations between the parties involved. However, Desjardins notes that information provided by the parties may not be accurate. Parties may not trust in the information provided.[35] The parties concerned, though, already have a minimal base of trust and respect for one another—Cold War Europe being an exception, as noted—so they are less likely to feed false information. Openness and transparency, as opposed to secrecy, may accelerate the process of confidence-building, as each party is more able to predict and understand the behavior of other parties.[36] Continued dialogue between the parties thus plays an invaluable role in confidence-building. CBMs, therefore, are not ends in themselves but are instruments used to build further confidence between certain parties.

As in the case of dialogue, therefore, which is after all a CBM, this confidence-building process must at some time start to address specific issues that could lead to definite agreements between the parties involved. These agreements could cover various military CBMs.[37] Negotiation only for the sake of negotiation will lead nowhere. Fred Tanner, in his analysis of the EU's interest in the Mediterranean, has suggested how this confidence-building process might lead to the introduction of specific measures in the Mediterranean or Euro-Mediterranean area. Following discussions and an exchange of information, there will be a need to work out certain negotiation principles and concepts. For example, the parties concerned need to agree on what is meant by expressions such as "defense sufficiency" or "legitimate defense requirements." Consensus would then need to be obtained on "codes of conduct," which would apply to the political-military sphere and also perhaps to military-civilian relations. These would provide the previously agreed-upon principles and concepts with an "operational dimension." For example, the prenotification of military exercises could be included in a code of conduct. In a next stage, certain "structural arrangements" could be decided. For instance, bodies could be created to handle the exchange of information and oversee verification procedures.[38] One may add that a crisis-prevention center could then be established. In reality, of course, a confidence-building process in the Mediterranean may not follow smoothly along these lines. It will be seen that in practice much more work is required in order to build further confidence in the Mediterranean or Euro-Mediterranean area.

It is worth noting that in the eastern Mediterranean, in the 1950s, 1960s, and 1970s, a number of limited subregional CBM regimes were concluded between Israel and Egypt, Israel and Jordan, and Israel and Syria. These regimes included such features as: certain provisions in armistice and peace agreements; the establishment of demilitarized zones; and the conclusion of "red-line arrangements" where geographical limitations were placed on the areas in which armies could operate, or where restrictions were imposed on military activities in certain locations.[39]

In the 1990s, within the framework of the Middle East peace process, the Working Group on Arms Control and Regional Security (ACRS) attempted to benefit from the CSCE experience and introduce various CBMs and CSBMs for the Middle East at a multilateral level. However, the activities of ACRS were suspended in 1995. This was because of the disagreement between Egypt and Israel over the issue of the proliferation of WMD in the Middle East.

Concerning CBMs and CSBMs or more advanced CBMs, it is important to take into account the differences between Cold War Europe and the post–Cold War Mediterranean area. How relevant is the past European experience for the Mediterranean of today? In Europe there was a single, major military stand off between NATO and the Warsaw Pact. The primary concern was to prevent either of the blocs launching a surprise large-scale attack. In the Mediterranean, on the other hand, there are massive assymetries in force structures between states north and south of the Mediterranean.[40] In the eastern Mediterranean the Israeli armed forces are clearly far superior than most of their Arab counterparts. Today's Mediterranean is more complex than yesterday's Cold

War Europe. Instead of two fixed military blocs there are shifting and impermanent alliances in the Mediterranean.⁴¹ In the eastern Mediterranean, in particular, there is a concern about a possible surprise attack by Israeli or Arab neighbours. However, it is extremely unlikely that in the present circumstances states north of the Mediterranean will launch a surprise attack against their southern neighbours, or vice versa. And, as already discussed, in contrast to Cold War Europe, Arab elites are not only concerned about military threats from other states. The economic, political, societal, and environmental aspects of security, which have an intrastate as well as interstate dimension, must be considered.

Nevertheless, some features of the CBMs and CSBMs of the Cold War period are also of relevance to the post–Cold War Mediterranean. For example, the ten principles of the Declaration of Principles of the Helsinki Final Act are of universal applicability. The Declaration noted that parties should: guarantee the sovereign equality of other states; refrain from the threat or use of force; respect the inviolability of state frontiers and the territorial integrity of states; solve disputes peacefully; avoid intervening in the internal affairs of other states; and respect human rights and fundamental freedoms. These principles apply to interstate and intrastate relations in the Mediterranean. The EU's Barcelona Process has emulated the Helsinki Final Act by creating three chapters or "baskets" basically covering military/political, economic, and societal issues.

It is worthwhile to consider how the new types of CBMs and CSBMs, which the CSCE and then OSCE have introduced in post–Cold War Europe, may be of relevance to the Mediterranean. It would seem unlikely that the Code of Conduct on Politico-Military Aspects of Security adopted at the CSCE/OSCE Budapest Summit in December 1994 could be applied wholesale to the Mediterranean. The Code of Conduct laid particular emphasis on the importance and need for the civilian control of military, paramilitary, and security forces within a state. This would not be possible to implement in many states in the southern and eastern Mediterranean where the armed forces have a dominant role in civilian affairs. At present, the Human Dimension mechanisms of the OSCE, with their emphasis on human rights and the value of free, open, and competitive elections, would also be impossible to implement in North Africa and the Middle East. Ruling Arab elites would likely complain that the Americans and Europeans were attempting to use the human rights issue as a pretext to interfere in the internal affairs of Arab states. Arab governments would be more open to initiatives that instead of challenging their internal power base, focus on addressing interstate problems. Therefore, for example, a conflict-prevention center could be created modeled on that established by the CSCE in 1990, and it could become a clearinghouse for the exchange of military information. And/or a Mediterranean version of the Forum for Security Cooperation, set up in 1992 within the then CSCE, could be formed. This Forum serves as an arena for the exchange of ideas and information on issues such as arms control, verification, and disarmament. The Forum also has developed a program that encourages contacts and cooperative activities between the armed forces of different states. Also organized by the Forum are joint military exercises, training programs, seminars, and workshops. Some of these features could be adopted to encourage cooperation between the armed forces of states in the southern and eastern Mediterranean—although Arab states will be unwilling to cooperate with Israel—and between the armed forces of states north and south of the Mediterranean. The importance of cooperative activities in the military field between NATO and the six non-NATO dialogue countries will be examined later in this study.

In the opinion of Heinz Vetschera, a number of lessons may be drawn from the CSCE experience in Europe and applied to the Mediterranean. The confidence-building process gradually evolved in Europe under the auspices of the CSCE. Dialogue between states resulted in agreement on certain common rules. The good will of the parties enabled the process to develop. In time it became possible to establish various mechanisms and permanent structures. The "Conference on" became an "Organization for" Security and Cooperation in Europe. A similar incremental process could perhaps work for the Mediterranean too. The CSCE experience demonstrated that in order to properly address regional security issues a comprehensive framework was required that included all relevant actors in Europe.[42]

One potential problem with the NATO-Mediterranean dialogue is that it involves only six non-NATO Mediterranean countries. Even the more inclusive Barcelona Process of the EU excludes the "rogue" states—Iran, Iraq, and Libya. One wonders, therefore, how far these initiatives can advance given their present formats.

Because of the current problems and tensions in the Mediterranean area as a whole it would seem that only over a considerable period of time would the introduction and implementation of certain CBMs and CSBMs lead to circumstances where a warm peace as opposed to a stable peace prevailed. This is in spite of the fact that most governments in the Mediterranean area are becoming increasingly aware of the costs of war and the risks of the escalation of tensions and are developing more interest in resolving economic, social, and ecological problems in the area. Employing other terminology, a comprehensive "security regime" applying to the Mediterranean in general is highly unlikely for the foreseeable future. The emergence of a so-called security community is an even more distant prospect.

Robert Jervis has defined a security regime as "those principles, rules and norms that permit nations to be restrained in their behavior in the belief that others will reciprocate. This concept implies . . . a form of cooperation that is more than the following of short-term self interest." Jervis listed the necessary conditions for forming and maintaining a security regime. Influential states should be supportive. These states would thus have to be relatively satisfied with the status quo. War must be regarded as too costly. Each state must have a shared interest in long-term cooperation in the belief that all will thereby benefit. This would prevent a state or states reneging on their implicit commitments.[43]

In a regime, cooperation over a long period on particular issues leads to predicability and a regularization of behavior. Regimes reduce uncertainty, as policies are coordinated, and the provision of information by this means enables negotiators to conclude mutually beneficial agreements.[44] As well as contributing to the possible formation of a security regime, CBMs and CSBMs would also be part of such a regime. A security regime would be likely to have a more developed institutional infrastructure to coordinate these CBMs and CSBMs. Particular norms of behavior that would dictate the policies of states would have emerged. International regimes require continued joint policy coordination.[45] Dialogue, therefore, would still be necessary, to help to sustain the regime, although in the process of time the parties concerned will become more familiar with each other. There would still be the fear, though, that one party might opt to defect from the regime in the belief that through this action it could secure advantages over the other parties in the regime.

Further cooperation between states on security issues could lead to the eventual formation of a "security community." In such a community there would be "real assurance that the members of that community will not fight each other physically but will settle their disputes in some other ways."[46] The term "security community" has been applied to developments in western Europe since the end of the Second World War. There is also talk now of fully incorporating central and eastern Europe into this security community after the end of the Cold War. In a security community, in which there are excellent channels of communication between the states involved, a sense of community evolves based on common values. This is strengthened and reinforced by shared institutions. None of the actors perceive each other as a threat. Relations between the members of a security community are based on "amity." In a security regime there remain elements of friction. There is still the possibility of conflict in a security regime. Relations between states in a security regime are based on "controlled enmity."[47] Genuine allies participate in a security community. They consult with one another and seek advice. Dialogue between members of a security community would not be required.

Obviously, it is far too early to discuss the prospects for states around the Mediterranean forming a security regime or a security community. The problems in relations between states north and south of the Mediterranean, between Arab states and Israel, and among the Arab states themselves, have already been noted. The internal politics of each state should also be considered. One cannot assume that the state is a monolithic entity.[48] In the case of the Mediterranean, therefore, the internal security problems of states in the southern and eastern Mediterranean and the different perceptions of ruling Arab elites and their publics must be considered. The relevance of the cultural factor for intrastate relations in Arab countries as well as for relations between Arab and non-Arab states must also be taken into account. The fact that elites in the West and in the Arab world seem to perceive security issues from dissimilar perspectives is also of significance.

In the current situation in the Mediterranean there is thus a need for dialogue between governments and peoples. A dialogue could take place between a government and people within the same state as well as between governments and peoples of different states. The NATO-Mediterranean dialogue aims to focus on relations between governments and different groups in states north and south of the Mediterranean. Its success or failure will be dependent on a whole range of factors that will be discussed in this study. It is important to remember, though, that the question of how the Middle East peace process evolves or does not evolve will be of crucial importance for security in the Mediterranean as a whole.

CONCLUSION

For NATO the concept of dialogue is nothing new. The term "dialogue" has been an important word in the NATO lexicon for some time. NATO's famous 1967 Harmel Report on Future Tasks of the Alliance, which was promulgated in a period of detente in East-West relations, stressed the importance of "dialogue" and "deterrence" vis-à-vis NATO's policy toward the Warsaw Pact. Dialogue could serve to improve East-West relations by reducing tensions and contributing to bringing about a more stable relationship.[49] The term "dialogue" was obviously employed in a totally different set of circumstances in 1967. At the time, there was a real military threat to the Atlantic Alliance from the Soviet-led Warsaw Pact. In the late 1990s the possible danger that NATO states may have to confront from the southern and eastern Mediterranean is in no way comparable with the past Soviet threat. Many commentators and analysts prefer to refer to challenges and risks rather than threats to the Atlantic Alliance from the south. But NATO officials believe that a dialogue with the south may help to reduce tensions and improve the security situation in the Mediterranean.

In line with other recent and ongoing initiatives concerned with security issues in the Mediterranean, the NATO-Mediterranean dialogue is an important tool of preventive diplomacy. The focus of the dialogue is more on diplomatic, political, and military concerns and less on social, economic, and cultural matters. NATO's Mediterranean Initiative is a CBM. The parties involved needed to have a certain amount of trust in one another in order to agree to participate in the Initiative. The NATO-Mediterranean dialogue should be further built upon. NATO's Mediterranean Initiative is in itself, therefore, a confidence-building *process* as well as *measure*. It is a process in which certain agreements have already been concluded and implemented with reference to a number of cooperative activities. Indeed, traditional and classic military CBMs are beginning to take shape within the NATO-Mediterranean dialogue.

The process is likely to be a long one. The NATO-Mediterranean dialogue is still in its early days. The membership of the dialogue remains limited in number. There are limits to what the dialogue may achieve as long as ruling Arab elites are primarily concerned with internal security problems. Further contacts between NATO officials and opinion leaders from the southern and eastern Mediterranean will help to overcome misperceptions and feelings of mistrust that are to an extent historically and culturally rooted.

Before focusing in more detail on current problems and issues in the Mediterranean area, it is important, first of all, to have a perspective on NATO's interests and involvement during the Mediterranean in the Cold War period. It will be seen that even before 1989 there were differences of opinion among NATO allies about what particular policies should be adopted in the southern and eastern Mediterranean. The ruling elites and peoples of the Arab states in the area formed certain impressions of the role of NATO in the Mediterranean in this period.

NOTES

[1] Peter Mangold, "Security: New Ideas, Old Ambiguities," *The World Today* 47, 2 (February 1991): 30–32.

[2] Barry Buzan, *People, States and Fear: The National Security Problem in International Relations* (Chapel Hill: University of North Carolina Press, 1983), 10.

[3] Helga Haftendorn, "The Security Puzzle: Theory-Building and Discipline-Building in International Security," *International Studies Quarterly* 35, 1 (March 1991): 3.

[4] Ken Booth and Peter Vale, "Security in Southern Africa: After Apartheid, beyond Realism," *International Affairs* 71, 2 (April 1995): 293.

[5] Mangold, "Security: New Ideas, Old Ambiguities," 30.

[6] Gareth Evans, "Cooperative Security and Intra-State Conflict," *Foreign Policy* 96 (fall 1994): 7.

[7] Barry Buzan, "Is International Security Possible?" in *New Thinking about Strategy and International Security*, Ken Booth, ed. (London: Harper Collins Academic, 1991), 35.

[8] Dieter Mahncke, *Parameters of European Security* (Paris: Chaillot Papers 10, Institute for Security Studies, WEU, 1993), 8.

[9] Jan Zielonka, "Europe's Security: A Great Confusion," *International Affairs* 67, 1 (January 1991): 128.

[10] Barry Buzan, "New Patterns of Global Security in the Twenty-First Century," *International Affairs* 67, 3 (July 1991): 447.

[11] Caroline Thomas, "New Directions in Thinking about Security in the Third World," in *New Thinking about Strategy and International Security*, 267–89.

[12] Mohammed Ayoob, *The Third World Security Predicament—State Making, Regional Conflict and the International System* (Boulder, Colo., London: Lynne Rienner, 1995), 42.

[13] Ibid., 178.

[14] Martin W. Sampson III, "Cultural Influences in Foreign Policy," in *New Directions in the Study of Foreign Policy*, Charles F. Hermann, Charles W. Kegley and James N. Rosenau, eds. (Boston: Allen and Unwin, 1987), 385–86.

[15] Samuel P. Huntington, "The Clash of Civilizations?" *Foreign Affairs* 72, 3 (summer 1993): 22–49.

[16] For several critiques of Huntington's article, see the various pieces in *Foreign Affairs* 72, 4 (September–October 1993).

[17] Kevin Avruch and Peter W. Black, "Conflict Resolution in International Settings: Problems and Prospects," in *Conflict Resolution Theory and Practice: Integration and Application,* Dennis J. D. Sandole and Hugo van der Merwe, eds. (Manchester, England, and New York: Manchester University Press, 1993), 132–34.

[18] Michael J. Mazarr, "Culture and International Relations: A Review Essay," *The Washington Quarterly* 19, 2 (spring 1996): 179–80.

[19] Ole Elgstrom, "National Culture and International Negotiations," *Cooperation and Conflict* 29, 3 (1994): 290.

[20] Alastair Ian Johnston, "Thinking about Strategic Culture," *International Security* 19 (spring 1995): 34.

[21] Robert D. Putnam, "Diplomacy and Domestic Politics: The Logic of Two-Level Games," *International Organization* 42, 3 (summer 1988): 427–60.

[22] Mazarr, "Culture and International Relations," 190–91.

[23] Hugh Roberts, "Algeria's Ruinous Impasse and the Honourable Way Out," *International Affairs* 71, 2 (April 1995): 255.

[24] John S. Murray, "Using Theory in Conflict Resolution Practice," in *Conflict Resolution Theory and Practice,* 229.

[25] Marie-France Desjardins, *Rethinking Confidence-Building Measures: Obstacles to Agreement and the Risks of Overselling the Process* (Oxford and New York: Adelphi Paper 307, Oxford University Press for the International Institute for Strategic Studies, 1996), 34–35.

[26] P. H. Gulliver, *Disputes and Negotiations: A Cross-Cultural Perspective* (New York: Academic Press, 1979), 70.

[27] Fred Tanner, "The Euro-Med Partnership: Prospects for Arms Limitations and Confidence-Building after Malta," *The International Spectator* 32, 2 (April–June 1997): 7.

[28] Desjardins, *Rethinking Confidence-Building Measures,* 5.

[29] Ibid.

[30] Ariel E. Levite and Emily B. Landau, "Confidence and Security-Building Measures in the Middle East," in *Regional Security in the Middle East: Past, Present and Future,* Zeev Maoz, ed. (London, and Portland, Oreg.: Frank Cass, 1997), 151, 155.

[31] Conference on Security and Cooperation in Europe (CSCE), Helsinki Final Act (Helsinki, 1975), Document on *Confidence-Building Measures and Certain Aspects of Security and Disarmament,* Preamble, paragraph 4, 84.

[32] Levite and Landau, "Confidence and Security-Building Measures in the Middle East," 148–49.

[33] Richard E. Darilek and Geoffrey Kemp, "Prospects for Confidence- and Security-Building Measures in the Middle East," in *Arms Control and*

Confidence-Building in the Middle East, Alan Platt, ed. (Washington, D.C.: US Institute of Peace, 1992), 9–42.

[34]Desjardins, *Rethinking Confidence-Building Measures,* 8–9, 12–13.

[35]Ibid., 18–19, 61–62.

[36]Presentation by Heinz Vetschera in a North Atlantic Assembly (NAA) Workshop on Confidence-Building Measures in the Mediterranean Region (Naples, April 11–12, 1997).

[37]Desjardins, *Rethinking Confidence-Building Measures,* 42, 61.

[38]Tanner, "The Euro-Med Partnership," 23–24.

[39]Yair Evron, "Confidence- and Security-Building Measures in the Arab-Israeli Context," in *Middle Eastern Security: Prospects for an Arms Control Regime,* Efraim Inbar and Shmuel Sandler, eds. (London, Portland, Oreg.: Frank Cass, 1995), 152–72.

[40]Darilek and Kemp, "Prospects for Confidence- and Security-Building Measures in the Middle East," 27–28.

[41]Gerald M. Steinberg, "European Security and the Middle East Peace Process," *Mediterranean Quarterly* 7, 1 (winter 1996): 72–73.

[42]Heinz Vetschera, "Regional Security and Arms Control—The CSCE Experience with Confidence-Building Measures and Crisis Mechanisms," Paper available at the Secretariat of the NAA, Brussels. See 24–25.

[43]Robert Jervis, "Security Regimes," in *International Regimes,* Stephen D. Krasner, ed. (Ithaca, N.Y., and London: Cornell University Press, 1983), 173, 176–81.

[44]Robert O. Keohane, *After Hegemony: Cooperation and Discord in the World Political Economy* (Princeton, N.J.: Princeton University Press, 1984), 97, 107.

[45]Etel Solingen, "The Domestic Sources of Regional Regimes: The Evolution of Nuclear Ambiguity in the Middle East," *International Studies Quarterly* 38, 2 (June 1994): 305–06.

[46]Karl W. Deutsch, *Political Community and the North Atlantic Area* (Princeton, N.J.: Princeton University Press, 1957), 5.

[47]Barry Buzan, "Introduction: The Changing Security Agenda in Europe," in *Identity, Migration, and the New Security Agenda in Europe,* Ole Waever et al., eds. (New York: St. Martin's, 1993), 10.

[48]Solingen, "The Domestic Sources of Regional Regimes," 305–07.

[49]For the text of the Harmel Report on Future Tasks of the Alliance issued by the Ministerial Sessions of the North Atlantic Council, the Defence Planning Committee, and the Nuclear Planning Group, Brussels, December 13–14, 1967, see *Texts of Final Communiqués 1949–74* (Brussels: NATO Information Service), 198–202.

CHAPTER 3
NATO and the Mediterranean in the Cold War

INTRODUCTION

As in international politics in general, the ending of the Cold War and the disintegration of the Soviet Union were a watershed with regard to developments in and around the Mediterranean. The principal focus of NATO in the Mediterranean had been to limit Soviet influence in the area. There was a competition between NATO and the Warsaw Pact to secure the support of governments in North Africa and the Middle East. In contrast, at present, in the post–Cold War period, notwithstanding Moscow's attempts to install S-300 missiles in Cyprus and then in Crete, Russia does not pose a serious challenge to NATO's interests in the Mediterranean.

One would not have a clear picture of events in and around the Mediterranean in the Cold War period, however, by exclusively analyzing processes and outcomes through an East-West prism. Indeed, for NATO members Spain and Portugal, for example, the Soviet threat was a distant one. The states of NATO's Southern Region—namely, France, Greece, Italy, Portugal, Spain, and Turkey—were watchful of developments in countries to their immediate south and east. There were tensions in north-south relations that the Soviet Union was not always able to exploit. As already noted, these tensions would result in NATO eventually launching its Mediterranean Initiative. And, as in the post–Cold War period, there were also problems between non-NATO states south and east of the Mediterranean, of which the Arab-Israeli conflict received most attention. Chapters 4 and 5 will examine these south-south issues and provide an overview of the political, economic, and social problems of North Africa and the Middle East. In the Cold War period there were also differences of opinion among NATO allies concerning how to respond to conflicts and crisis situations in and around the Mediterranean. These intra-alliance divisions arguably still exist today, as south-south and north-south issues have assumed more prominence in their own right after the ending of East-West tensions.

There is a tendency among NATO officials in recent years, though, to view the Mediterranean area more as a whole. Previously, the eastern and western parts of the Mediterranean were regarded almost as separate zones. The former was connected with problems in the Middle East, the Gulf, and the Aegean where the focus was on military, defense matters, the Arab-Israeli question, and the oil issue. In particular, Italy and Turkey perceived the Soviet threat as real. The United States tended to be less concerned with events in the western Mediterranean. In general, the European members of NATO were more interested in economic, social, and political problems such as the migration issue, which stemmed from the Maghreb. For countries in the western Mediterranean, the Soviet threat was much less immediate. As recently as 1990, France was lobbying for the permanent establishment of a Western Mediterranean Forum that would have as its members Italy, Portugal, Spain, Malta, Algeria, Libya, Mauritania, Morocco, and Tunisia, in addition to France. However, after the end of the Cold War, this "bifurcated view" of the Mediterranean and Mediterranean security was beginning to erode. The US and European members of NATO became increasingly more aware of the overlap between the issues and problems of the eastern and western Mediterranean.[1] The distinction between the eastern and western Mediterranean in the eyes of Western officials may be traced back to the onset of the Cold War.

The aim of this chapter is to provide an overview of the policy of NATO and individual NATO member countries toward the Mediterranean in the Cold War period. In addition to the perceived Soviet threat, the so-called out-of-area issue is examined with reference to the Mediterranean. The out-of-area problem involved the Soviet threat and also north-south and south-south issues. It also drew attention to the close linkage between developments in the Mediterranean and in the Persian Gulf. On account of their various particular interests, individual NATO member countries held different interpretations of the significance of certain events that took place out-of-area. Divisions within the Atlantic Alliance were exposed. With the end of the Cold War, the distinction between events in-area and out-of-area began to blur. The

Mediterranean Initiative covers territory that NATO officials have regarded as out-of-area.

NATO, THE MEDITERRANEAN, AND THE COLD WAR

> *The Mediterranean has historically been an area of conflict, a hotbed of crisis, and a theatre of war.*
> —ANTONIO BADINI, "EFFORTS AT MEDITERRANEAN COOPERATION"

Throughout the Cold War many NATO officials sought to check Soviet ambitions in the Mediterranean. Certainly, Moscow attempted to gain political influence in states such as Algeria, Egypt, Libya, and Syria. The Atlantic Alliance was keen to prevent the Soviet navy from securing access to and use of naval bases in the area. There was a fear that the Soviet fleet could threaten lines of communication and disrupt trade routes. US officials were particularly determined to maintain access to Israel, a key strategic ally of the United States in the eastern Mediterranean. Partly because of the Mediterranean–Persian Gulf connection and the transport of oil, the Mediterranean was "the most intensely utilized maritime corridor of the world."[2] The Europeans were more dependent than the United States on energy supplies from the Persian Gulf and the Mediterranean. The continued safety of the passage of tankers and pipeline connections from the Persian Gulf, the Middle East, and North Africa was uppermost in the thinking of many NATO policy-makers. In 1989 North Africa alone satisfied 20.5 percent of Western Europe's gas consumption needs, the Middle East supplied 31.5 percent of European oil demand, and North Africa provided a further 15.4 percent of the 12.51 million barrels of oil consumed each day in Europe.[3]

In spite of these causes for apprehension, the Mediterranean was never as important for NATO as the so-called Central Front. Most NATO forces were concentrated in central Europe to counter an anticipated massive Warsaw Pact assault. It has been argued that Atlantic Alliance officials in the Cold War regarded Allied Forces Southern Europe (AFSOUTH) based in Naples, one of the regional subordinate commands of NATO's Allied Command Europe (ACE), "only as an adjunct to strategic considerations in Central Europe."[4] France and Spain were not integrated into NATO's military structure. NATO member states in the Mediterranean, most notably Greece and Turkey, were not always on the best of terms.

As the Second World War drew to a close, Moscow pressured its Western wartime allies to give the Soviet Union territorial concessions in the Mediterranean. The Soviets demanded trusteeship over the former Italian colony of Tripolitania (Libya) and the right to establish a military base on Turkish territory by the Straits. Kremlin officials then sought to secure joint control of the Straits with the Turkish authorities. At the same time, farther east, the Soviet Union was attempting to consolidate its presence in Iran. In the face of these pressures in spring 1946 the United States dispatched the warship the USS *Missouri* to Istanbul. This was the first post–Second World War reinforcement of US forces in Europe. Other US warships were soon deployed in the area, leading to the eventual formation of the US Sixth Fleet. Because Britain was no longer equipped with the resources to maintain a substantial presence and influence in the Mediterranean, it was left to the United States to pledge continued economic and financial aid—and also, thereby, military support—to Turkey and Greece, through the announcement of the Truman Doctrine in March 1947. Alarmed at Soviet demands, by the spring of 1947 Turkish policy-makers were lobbying for the creation of a Western-supported Eastern Mediterranean Pact composed of Turkey, Greece, and Egypt. An Eastern Mediterranean Pact could work closely with a proposed Western Mediterranean Pact of France, Spain, and Italy.[5] Later in the same year Turkish officials promoted the idea of a single Mediterranean Security Pact that would also include the United States and Britain.[6] However, this attempt to bridge what Britain perceived to be two distinct eastern and western Mediterranean zones did not succeed.

In July 1948 in Washington, D.C., the lengthy preparatory talks commenced that would finally culminate in the signing of the North Atlantic Treaty on April 4, 1949. In these talks Greece and Turkey were not seriously considered potential signatories of the treaty. They were not invited to participate in the discussions. Once the talks were under way the United States, Canada, Britain, France, and the Benelux countries invited Portugal to attend, largely because of Portugal's possession of the strategically important Azores in the Atlantic Ocean. Lively discussions followed about which other states should be a party to the treaty and over what territories, should a possible security guarantee apply—the assumption being that the threat to signatories would come from the Soviet Union. France clamored for extending a security guarantee to cover all its North African possessions, namely Algeria, Morocco, and Tunisia.

Other participants in the talks rejected the French move on the grounds that it would divert attention from the central issue of defending western Europe. The French delegation was still able to incorporate the Algerian departments of France in the eventual North Atlantic Treaty. Here, an exception was made to the principle that no colonial territories should directly benefit from any security guarantee. The United States had been adamant on this issue in the face of attempts by Britain, the Netherlands, and Portugal to obtain defense guarantees for their colonies. France also successfully lobbied for Italy to be admitted as a signatory. Only then would France support the admission of any Scandinavian countries. Other participants were willing to concede to French proposals partly in the hope of preventing the Communist parties from securing more electoral support in France and Italy.[7]

When the North Atlantic Treaty was signed in 1949, a "minimal southern flank" had been reluctantly agreed to "based primarily upon political compromises rather than geostrategic logic."[8] Nevertheless, a western and central Mediterranean component was in effect added to the North Atlantic Treaty. After the treaty's signing, for a few more years, the French government would continue to press for greater attention to the Mediterranean and North Africa because the Second World War had demonstrated that North Africa and Europe were actually a single theater of operations. The US administration suspected that French officials were determined to cling to their North African colonies by whatever means.[9]

As the signatories of the North Atlantic Treaty quickly moved to establish a treaty organization in the wake of the outbreak of the Korean War, Britain sought to keep the eastern Mediterranean excluded. The objective of the British government was to establish a separate Mediterranean Pact in which Britain would participate together with Turkey and Egypt and possibly Greece. This Mediterranean Pact could have helped Britain to maintain its control over disputed bases along the Suez Canal.[10] By October 1951 Egypt had denounced the plans to set up a Mediterranean Pact. Britain remained committed to creating a Middle East Command and then a Middle East Defense Organization but these projects also fell victim to Arab hostility. Eventually, what turned out to be the largely ineffectual Central Treaty Organization (CENTO) would instead emerge. In the meantime, as hostilities in Korea escalated, in September 1950 Greece and Turkey were allowed to be associate members of NATO—a status that did not provide automatic defense guarantees. In September 1951 at the Ottawa meeting of NATO foreign ministers it was finally decided to admit Turkey and Greece into the Atlantic Alliance. In February 1952 the two states acceded as full members. The first enlargement of NATO had strengthened the Mediterranean component of the Atlantic Alliance by extending the defense guarantee to the eastern Mediterranean.

The Suez Crisis of 1956, sparked by the intention of the Egyptian government under Colonel Gamal Abdel Nasser to nationalize the Suez Canal, was significant for various reasons. An open rift in NATO was exposed between the French and British on one hand and the United States on the other. As will be discussed later in this chapter, the crisis had major repercussions for the so-called out-of-area question. The growing Soviet influence in the Mediterranean was revealed with Moscow's provision of weaponry and economic and diplomatic support for the Nasser government. The Soviet Union was seeking to present itself as the champion of the forces of decolonization and anti-Western imperialism. The Suez Crisis was an example of how East-West and north-south issues could intertwine in this period.[11] Soviet naval forces established a permanent presence in the Mediterranean with the appearance in 1964 of the Fifth Eskadra, a detachment of the Soviet Black Sea fleet. The mission of this naval force was to show the flag and attempt to counter US ballistic missile submarines and cruisers active in the Mediterranean Sea.[12]

By 1973 the Soviet naval presence in the Mediterranean was at its peak. On average each day fifty-eight vessels were operational in the area.[13] In the Arab-Israeli War of that year the Soviet threat to intervene with airborne troops forced the United States to place its forces on a nuclear alert for a brief period.

Unlike the US Sixth Fleet, the Soviet Union's Fifth Eskadra failed to obtain full base rights in a Mediterranean country apart from access to and permanent port facilities in the Egyptian port of Alexandria until 1972. The Soviet naval threat diminished as a consequence. Only berthing facilities were provided at Benghazi and Tripoli in Libya, and at Mers-el-Kebir in Algeria, while bunkering facilities were offered off Malta.[14] Limited port facilities were also provided in Latakia in Syria. After 1973 the Soviet naval presence in the Mediterranean declined. Moscow continued to transfer arms exports and supply military technicians to Algeria and Libya, though, almost up to the break up of the Soviet Union.

In response to the Soviet Union's growing interest in the Mediterranean, in 1961 NATO set up an Expert Working Group on the Middle East and the Maghreb. Six years later an Ad Hoc Group on the Mediterranean, composed of officials from the national capitals of member states, was formed. Both groups only met twice for two days each year in Brussels. After 1986 discussions conducted in these groups no longer solely focused on the East-West context. These groups began to examine general developments around the Mediterranean. They also analyzed topics such as the crises in Yugoslavia and Algeria, problems in Yemen, Libya, and Iraq, north-south issues, and the Middle East peace process.[15] In the words of a report of the NAA, neither group "has proved as active, informed or forward-looking as events warrant."[16]

With the Soviet naval expansion of the mid-1960s, and following the recommendations of the NAC in Reykjavík in June 1968, NATO decided to establish a Maritime Air Force in Naples to coordinate surveillance in the Mediterranean.[17] In 1969 NATO also agreed to deploy a naval-on-call force for the Mediterranean, NAVOCFORMED.

The Soviet naval presence in the Mediterranean may have been declining, but with the invasion of Afghanistan in 1979 the US administration, in particular, was alarmed that the Soviet Union might be tempted to strike at Iran in order to gain control of the Persian Gulf. In that eventuality, Turkey would have felt increasingly threatened, and Western commercial interests would have suffered a devastating blow. The Gulf War between Iran and Iraq in the 1980s actually did seriously jeopardize Western access to the Persian Gulf. NATO member countries came to realize that south-south problems, in addition to the perceived Soviet threat, could have a major negative impact on their economic interests. As noted in the following section, these developments in Afghanistan and in the Persian Gulf did have potential and actual consequences for NATO's defense concerns in the Mediterranean.

Not all NATO member states shared to the same extent the US fears and concerns about the Soviet Union. Some governments regarded the Soviet threat as remote, and this encouraged the distinct and at times assertive national policies of some allies. In the 1970s and 1980s some NATO member states in the Southern Region perceived that their security concerns were not properly met by the Atlantic Alliance, in part because of what they regarded as a US obsession with the Soviet threat. Indeed, NATO member states in the Mediterranean at times were at complete odds with the United States. In the 1970s Greece withdrew temporarily from NATO's integrated military structure because of what Athens regarded as a pro-Turkish stance by the United States at the time of the conflict in Cyprus. Greek officials believed that the US administration was lenient toward Ankara due to Turkey's strategic value with regard to the Soviet Union.[18] Turkish-American relations also nosedived, though, when the US Congress slapped an arms embargo on Turkey between 1975 and 1978 following hostilities in Cyprus. Sensitive to their public opinion, Spanish officials forced the withdrawal of United States forces from Spain in the late 1980s. Italy and the US clashed in 1985 over the handling of the *Achille Lauro* affair when the Italians allowed a Palestinian terrorist to escape.[19] From the mid-1960s onward France periodically complained of the size and nature of the United States naval presence in the Mediterranean.

One analyst declared at the time of the Cold War that the principal challenge NATO faced in the Southern Region did not stem from the Soviet Union or other non-NATO states in the area. The main challenge was rather one of internal management. This was because of the different political interests and national concerns of member states in a vast territory stretching from the Azores to Ardahan in eastern Turkey.[20] As well as difficulties between the United States and allies in the Southern Region, there were also strains between other NATO members. The numerous problems in Turkish-Greek relations are well known. There is also a traditional rivalry between Portugal and Spain. Officials in Lisbon were concerned that if Spain were to be fully integrated into NATO's military command structure, Portuguese forces would be placed under direct Spanish control. That would be viewed, in Portugal, as a highly unpopular measure.

At the height of the Cold War, with the exception of France, NATO member states in the Southern Region were willing to play the role of "good allies" of the United States. But as the NATO Mediterranean states developed politically and economically, and as some of them became less apprehensive of Soviet activities, there was an increasing "maturation" of the foreign policies of southern European states. Their interests began to diverge from those of the United States. By the 1980s the southern Europeans had rediscovered their "Mediterranean vocation" and were becoming more attracted toward developments in North Africa and the Middle East. Italy was at the forefront.[21] Northern European members of NATO were less interested in the politics of the Mediterranean. These allies would later welcome the Mediterranean Initiative with little enthusiasm. In the 1980s, there was a growing awareness among the southern Europeans that threats to their security were more likely to come from crises and conflicts beyond their southern borders than from an East-West confrontation.[22] This was seen in the responses of different NATO members to the out-of-area debate.

NATO, OUT-OF-AREA, AND THE MEDITERRANEAN

The term "out-of-area" as opposed to "in-area" was used in NATO circles with reference to Articles 5 and 6 of the North Atlantic Treaty signed in Washington, D.C., in 1949. The focus was on the security guarantees offered by NATO. The Atlantic Alliance sought to deter an attack against one of its members. An aggressor could face a massive retaliatory response.[23] An armed attack against one or more members would be considered by NATO an attack against all members of the Alliance—if the attack was against the territory of a party to the treaty in Europe or North America, or was against the forces, vessels, and aircraft of a party in the Mediterranean or in the North Atlantic above the Tropic of Cancer. The treaty also initially included the French territory of Algeria in this context, thereby making the Mediterranean Sea and a part of North Africa "in-area." The security guarantee of NATO would not automatically apply to other territories—such as the colonies of the signatories—which were referred to as "out-of-area." Nevertheless, by Article 4 of the North Atlantic Treaty signatories could consult on security concerns stemming from threats out-of-area. The treaty itself did not prohibit NATO members from individually or collectively engaging in operations out-of-area. Article 2 of the North Atlantic Treaty noted that all parties were obliged to contribute to peaceful and friendly international relations by "promoting conditions of stability and well-being." Throughout the Cold War, the out-of-area debate tended to concentrate on whether NATO would participate in out-of-area military operations to protect the interests of Atlantic Alliance members and most likely thwart Soviet ambitions. There was much less talk of NATO possibly becoming involved in nonmilitary activities out-of-area.

The out-of-area question assumed more prominence in the immediate aftermath of the Suez debacle. Atlantic Alliance members grew more apprehensive about the Soviet Union possibly further exploiting Arab-Israeli tensions to secure an advantage in the eastern Mediterranean. Britain and France had been involved in a military operation out-of-area without consulting, let alone seeking, the approval of other NATO members beforehand. Article 4 of the North Atlantic Treaty had been ignored. In December 1956 NATO published the so-called Report of the Three Wise Men. The Report cautioned: "NATO should not forget that the influence and interests of its members are not confined to the area covered by the Treaty, and that common interests of the Atlantic Community can be seriously affected by developments outside the treaty area."[24] No specific recommendations were made, though, concerning how NATO should respond to certain developments out-of-area. The Atlantic Alliance was similarly vague about how it should react to Soviet behavior in the Middle East. The communiqué of the NAC, issued in Paris in December 1956, noted:

> The Council discussed the threat which Soviet penetration into the Middle East could present for NATO. In view of the fact that the security, stability and well-being of this area are essential for the maintenance of world peace, the Council agreed to keep developments in this area under close and continuing observation.[25]

In the late 1950s, French President Charles de Gaulle proposed that NATO's jurisdiction be extended to include North Africa and the Middle East in accordance with his plan for the United States, Britain, and France to manage the Atlantic Alliance as a tripartite directorate. This may have been a resurrection of the earlier French demand of the late 1940s in the Washington preparatory talks to include in effect "in-area" what were then the extensive French colonial territories in North Africa. It seems that de Gaulle's suggestion was partly connected with the problem of Algeria—which by that time was engaged in its war of independence. De Gaulle, though, was also apparently attempting to steer France toward a more autonomous course of action within NATO.[26] Failing to secure the support of other allies, in March 1959 de Gaulle announced that France was withdrawing its Mediterranean fleet from NATO command. Seven years later all French forces pulled out from NATO's integrated military command.

The famous Harmel Report released in 1967 on the Future Tasks of the Alliance noted the worsening situation in the Middle East and stated that crises and conflicts outside the North Atlantic Treaty area could affect the security of NATO. The Report continued: "In accordance with established usage the Allies or such of them as wish to do so will also continue to consult on such problems without commitment and as the case may demand."[27] Again, the expressions used were vague and noncommittal. It was by no means clear what was "established usage" concerning consultation between allies on out-of-area issues. In the case of the eastern Mediterranean, in 1958 the United States and Britain had militarily intervened in Lebanon and Jordan, respectively, to shore up local governments from possible Soviet encroachment, without consulting other allies beforehand. And in the 1973 Arab-Israeli War, there was a lack of consultation when the United States placed its forces on nuclear alert to forestall possible Soviet military involvement.

The out-of-area issue assumed much more significance for NATO in the 1980s after the fall of the Shah of Iran and the Soviet invasion of Afghanistan. The US administration feared Moscow could attempt to push south through a destabilized Iran in order to throttle Western oil supplies coming from the Persian Gulf. In January 1980, US President Jimmy Carter announced the establishment of the Rapid Deployment Force (RDF), which could be used to defend the Persian Gulf from outside—Soviet—intervention. Other allies were not consulted even though the dispatch of the RDF would have meant a 20 to 33 percent shortfall in the number of US troops available to reinforce Europe in a crisis.[28] Most of the European members of NATO did not share American fears of the Soviet Union. They were reluctant to destroy detente and jeopardize major commercial deals with Moscow. There was a widely held belief that Soviet behavior could be tempered through continued economic cooperation. Many in the United States were critical of the Europeans' reluctance to support wholeheartedly the United States, especially since Europe was much more dependent on energy supplies from the Persian Gulf. However, there would be more unity within the Atlantic Alliance concerning out-of-area issues in the Gulf and in the eastern Mediterranean once the immediate Soviet threat diminished and the Iran-Iraq War of the 1980s intensified.

At the outbreak of the Iran-Iraq War, naval forces from the United States, Britain, and France coordinated their activities at the operational level in the Indian Ocean in order to keep open the Straits of Hormuz to commercial traffic. In 1984, maritime units from the United States, Britain, France, Italy, and the Netherlands were dispatched on mine-hunting operations in the Red Sea. Three years later naval vessels from these same countries were deployed in the Gulf to press for a cease-fire in the Iran-Iraq War. With regard to the eastern Mediterranean specifically, in 1981 the United States, Britain, France, Italy, and the Netherlands provided contingents for the Multinational Forces and Observers based in the Sinai Desert. After the Israeli attack on Beirut in 1982, a Multinational Force of American, French, Italian, and British troops was deployed in Lebanon for two years. These out-of-area operations were basically preventive military measures. They were coordinated at the operational level but were not sanctioned by NATO. However, as will be noted shortly, the naval deployments in the Gulf did necessitate a readjustment of NATO forces in the Mediterranean. As early as September

1983, the director of the International Institute for Strategic Studies in London proclaimed—perhaps somewhat surprisingly—that NATO and its relationship with the Third World was the most important issue that the Atlantic Alliance confronted.[29] Certainly, more NATO member states were becoming interested in events out-of-area. It has even been suggested that this interest of the Europeans was only in part due to the need to keep open trade routes, as there was also ostensibly a concern to restrain the United States in the Mediterranean and the Gulf.[30]

The European members of NATO were not willing to promise military support to the RDF. They were aware, though, that the Atlantic Alliance should at least engage in a more extensive discussion on out-of-area issues in order to pacify many in the US Congress and much of the American public who believed that Europe should be sharing more of the burden with NATO. Thus, in 1980 and 1981 official NATO communiqués referred to out-of-area problems, and questions were raised about the possible deployment of the RDF. Eventually, the so-called compensation-consultation-facilitation formula was agreed on by the Atlantic Alliance at the Bonn Summit in 1982. It was noted that as developments beyond the NATO area might threaten vital interests of the Atlantic Alliance, there was a need for allies to "consult" on these issues and "examine the requirements which may arise for the defence of the NATO area as a result of deployment by individual members outside that area"—there was a need to "compensate" for forces deployed out-of-area. The Atlantic Alliance also agreed to look into steps individual allies might take "to facilitate possible military deployments beyond the NATO area" that could be an important contribution to Western security.[31] This formula has been subjected to criticism. According to one commentator, it was imprecise, open to ad hoc interpretation, and achieved little concrete regarding an out-of-area strategy. The Atlantic Alliance had supposedly produced a "suboptimal outcome which, although lacking a strategic rationale, met the test of political acceptability."[32] In practice, the record was a mixed one. For example, the United States only consulted a few allies before its air raid on Libya in 1986, while France moved forces into Chad in the 1980s to check Gadhafi's ambitions, without consulting any NATO member. Britain did facilitate the American strike on Libya by allowing bombers to operate from British bases, but France, Italy, and others denied the United States overflight rights. As for compensation, earlier in 1980 Italy had deployed more naval units in the Mediterranean when other allies dispatched vessels to the Indian Ocean. Later, the Federal Republic of Germany would take up the slack in the North Sea and in the

Mediterranean when more allied ships were ordered to the Gulf. In these cases, the intention was to ensure that the Soviet Union would not benefit from a possible reduced NATO presence in the Mediterranean. The close connection between the Persian Gulf and the Mediterranean was again evident.

The United States and many European members of NATO also differed in their approach toward the Arab-Israeli question. For US administrations, ever anxious about Soviet meddling in the Middle East, Israel was a vital strategic ally in the eastern Mediterranean. Most Europeans—more familiar than the United States with the Arab world, and more economically dependent on energy-rich Arab states—were in favor of a more balanced policy toward the Middle East and became increasingly critical of what appeared to be almost automatic American approval of every Israeli action. Within NATO only Portugal facilitated US support of Israel in the 1973 Arab-Israeli War. After the war and the oil crisis, the European Community (EC) initiated with members of the Arab League the so-called Euro-Arab Dialogue in 1974.

This first attempt at developing a genuine dialogue between states north and south of the Mediterranean was not very successful. The aim of the Euro-Arab dialogue was to improve relations in the political, economic, cultural, and technical fields. It was hoped that a comprehensive forum would thereby be established. A formal Euro-Arab Consultative Council was created by mid-1974 composed of the nine member states of the then EC and the twenty-one member states of the Arab League.[33] Various working groups were formed in 1975 and several meetings were held within the Euro-Arab dialogue between 1976 and 1978. After the conclusion of the US-sponsored Camp David Agreements, which concentrated on relations between Egypt and Israel, the Arab League requested the suspension of the dialogue. Attempts to restart the dialogue in December 1983 and June 1988 failed. In a speech before the European Parliament in 1993, President Zine el Abidine Ben Ali of Tunisia appealed in vain for a revival of the Euro-Arab dialogue.[34]

The Euro-Arab dialogue failed for a number of reasons. It was a part of the so-called European Political Cooperation (EPC). The EPC aimed to coordinate the foreign policies of the member states of the EC. The Europeans, here, hoped to differentiate more clearly their policy from that of the United States. However, in practice, the EPC was slow and cumbersome. Unanimity was required before any policy could be pursued. There was thus a lack of forward planning within the EPC.[35] The divisions within the member states of the EC were demonstrated in the inability of the EPC in general to establish a consensus in order to implement certain measures. This weakness impaired the workings of the Euro-Arab dialogue.

There were also serious divisions in the Arab camp after the outbreak of the Iran-Iraq War in 1980. Overall, the circumstances were not conducive for the launching of a genuine dialogue. The objectives of the Euro-Arab dialogue needed to be more sharply defined. The Europeans tended to focus on economic issues. The Arabs were more eager to discuss political matters and, in particular, the question of Palestinian self-determination.[36] In short, the Euro-Arab dialogue was bound to fail because of the serious differences of opinion between member states of the EC and the Arab League. The parties concerned were not able to agree on the type and range of issues to be discussed. The dialogue also failed to adopt regular procedures. Meetings tended to be haphazard.

However, arguably, there was a positive offshoot of the Euro-Arab dialogue that went some way toward building confidence between European and Arab states in the Mediterranean. In marked contrast to the recently concluded Camp David Agreements, the EC in its Venice Declaration in January 1980, supporting UN Security Council Resolutions 242 and 338, endorsed the call for full Palestinian self-determination, and agreed that the PLO should be associated with any peace negotiations. This was widely interpreted as a European recognition of the PLO. Thus, the United States and Israel opposed the Declaration. The United States only made efforts to hold discussions with the PLO in late 1988 after the PLO leadership renounced terrorism and recognized the right of Israel to exist.

In the Cold War period, direct "threats from the south" did exist in the form of terrorism and Gadhafi's Libya. In 1985, for example, in addition to the *Achille Lauro* affair, Palestinian terrorists massacred innocent civilians at Rome's international airport. In spite of US-Italian friction over the *Achille Lauro* affair, there was a general consensus within NATO that condemned acts of terrorism. The Libyan claim that the Gulf of Sidra was part of its territorial waters threatened freedom of navigation in the Mediterranean. In April 1986, following the American air strike against Libya after the United States suspected Libyan involvement in the bombing of a nightclub in Berlin frequented by American servicemen, Gadhafi ordered the firing of two SCUD missiles at the US-manned Long-Range Aids to Navigation station on the Italian island of Lampedusa. Although the missiles were well short of their intended target, the Italians could have called for the activation of the Atlantic Alliance defense guarantee against Libya. However, the government in Rome was against retaliation, in part because of its economic interests in Libya and the presence of Italian nationals there. Nevertheless, Italy reinforced its Mediterranean defense forces. The Italians became more concerned about other future possible threats out-of-area from the south.[37] Ironically, the previous year, the Italian government had established its own rapid reaction force, which could be deployed to evacuate Italian nationals working abroad in places such as Libya.

France and Tunisia concluded a security agreement in order to deter a Libyan attack against Tunisian territory. In 1980 a group of opponents to the Tunisian government who had been trained in Libya mounted an attack against the Tunisian border town of Gafsa. France immediately responded by deploying a fleet just beyond Libyan territorial waters and thereby forestalled a possible larger-scale Libyan offensive. By this time French officials were keen to improve relations with the states of the Maghreb and especially with Algeria. In January 1983 President François Mitterrand proposed a conference of western Mediterranean states that would include France, Italy, Spain, Morocco, Algeria, Tunisia, and possibly Portugal and even Greece. Plans for the conference were scuttled because of Algeria's reluctance to attend. France continued to maintain close bilateral relations with states of the Maghreb.[38] Mitterrand would finally succeed in assembling western Mediterranean states several years later. As briefly noted earlier, French forces did directly clash with Gadhafi's troops south of the Mediterranean in Chad in the mid-1980s. By that time the French had created their own rapid reaction force, which could take part in out-of-area operations.

One potential out-of-area problem in the Cold War period that could have posed problems for the Atlantic Alliance after the admission of Spain in 1982 involved the Spanish enclaves of Ceuta and Melilla in northern Morocco. The government in Madrid was worried about a possible repeat of the Green March, when thousands of Moroccans had literally marched into the Spanish Sahara (which became known as the Western Sahara) to claim that territory for Morocco. Ominously at the time, in 1984 Morocco and Libya signed a defense agreement. King Hassan of Morocco did occasionally raise the question of the enclaves. It seemed that he was using this as a bargaining chip to secure further economic aid from Spain, Spanish support in Morocco's dealings with the EC, and the sympathy of Madrid with regard to the Western Saharan issue. At no time did the king threaten to attack the enclaves.[39] Spain concluded defense and cooperation agreements with Morocco, Tunisia, and Mauritania in the 1980s. Other NATO members negotiated bilateral defense agreements with southern Mediterranean states. France and the United States had signed such treaties with Algeria, Morocco, and Tunisia. Britain, France, and the United States have concluded similar defense agreements with Egypt. Although most of these agreements were primarily aimed at curtailing Soviet influence in the Mediterranean, the European members of the Atlantic Alliance, in particular, were also interested in developing ties with North Africa for their own economic and political interests. Turkey, meanwhile, was suspicious of the activities of its neighbors to the east and south. Officials in Ankara accused the authorities in Iran, Iraq, and Syria of lending assistance to Kurdish insurgents who were seeking to destroy the territorial integrity of the Turkish state.

At the height of the Cold War, there had probably been apprehension in Atlantic Alliance circles that coordinated NATO operations out-of-area might have triggered a Soviet military response. In practice, throughout the Cold War, NATO as a collective alliance did not participate in any out-of-area operation. At most, cooperation was limited to coordinated naval activities at the operational level between a handful of allies, or occasional half-hearted applications of the so-called compensation-consultation-facilitation formula. Allies differed over their perception of possible threats out-of-area, with the United States at times seemingly obsessed with the Soviet menace. In the words of an NAA Report on the out-of-area issue published in 1985, NATO as an alliance was not designed, it seemed, for out-of-area conflicts—"The security of the Allies does extend boundaries beyond NATO, but serious threats out-of-area are best viewed from a non-NATO perspective."[40] In 1988 a report produced by a working group of the Atlantic Council of the United States on the future of NATO concluded that the Atlantic Alliance as an organization, distinct from individual members, cannot and should not be a forum for action on out-of-area issues. In such cases the Atlantic Alliance would likely be "unsuccessful." Instead, to prevent further questioning in the United States on the continuing value and relevance of NATO, the report recommended that "the US and its NATO allies must pursue more active bilateral and multilateral consultations on issues outside the traditional Alliance purview."[41]

Throughout the Cold War, there was in general an assumption that the out-of-area debate concerned threats that might require emergency, preventive, or retaliatory military measures. The military/defense dimension of security was predominant in the case of either a Soviet threat or a challenge posed by other states in the southern and eastern Mediterranean and the Gulf. Nevertheless, in the 1980s there was a gradual awareness among NATO officials that Article 2 of the North Atlantic Treaty noted that signatories should not exclusively concern themselves with defense issues. The June 1982 Bonn Summit Declaration, in its references to out-of-area, spoke of the need for allies "to remove the causes of instability such as underdevelopment or tensions which encourage outside interference." There was also a promise to continue to struggle against global hunger and poverty.[42] The Declaration appears to have reflected the concerns of European members of the Atlantic Alliance. These members argued that political and other solutions to out-of-area problems should be pursued with recourse to military measures only as a last resort. The 1986 NAA Report on out-of-area issues noted that the Atlantic Alliance had gradually adopted a broadened concept of security. It was recognized that political, economic, and military aspects of security were interrelated. The Report added that allies were already offering political and diplomatic support, economic assistance, cultural cooperation programs, and military aid to non-NATO members. It recommended that these various forms of assistance be better coordinated.[43] In May 1989, when the Cold War was basically over, the communiqué of the NATO Summit in Brussels declared:

> In the spirit of Article 2 of the Washington Treaty, we will increasingly need to address worldwide problems which have a bearing on our security, particularly environmental degradation, resource conflict and grave economic disparities. We will seek to do so in the appropriate multilateral fora, in the widest possible cooperation with other states.[44]

Clearly, the out-of-area debate was no longer to be exclusively focused on potential military operations. Nevertheless, it did not seem that NATO itself was being seriously considered as a possible forum in which an out-of-area policy could be designed.

CONCLUSION

In the Cold War period, therefore, the United States and the Soviet Union competed to secure the support of Arab states in the Mediterranean. Even the Arab-Israeli conflict was to an extent subsumed within the wider East-West rivalry. Israel and Jordan were firmly in the Western camp, while Syria and various Palestinian factions leaned toward the Soviet bloc. Egypt would switch allegiances from East to West in the 1970s. In spite of the existence of the nonaligned movement, throughout the Cold War the Mediterranean was basically divided into those states and groups that supported the West—a definite majority—and those that favored the Soviet Union. Most Arab governments were fully aware that compared with the Soviet Union and its allies, the West was able to lend them more economic, political, and military assistance. In these circumstances, there was certainly no prospect for the Mediterranean states themselves to engage in region-building and promoting the development of a Mediterranean region.

The attempted Euro-Arab dialogue was significant for a number of reasons. In spite of Cold War tensions the Europeans made a concerted effort to further improve relations with Arab states. The EC was prepared to challenge the traditional US policy in the Middle East, which had been based on almost unquestioning support of Israel. One may argue that the EU's Barcelona Process has been able to benefit from this earlier exercise in European-Arab diplomacy. In contrast, because of Arab perceptions of NATO as a US-dominated organization, the Atlantic Alliance in its later Mediterranean Initiative has found it more difficult to secure trust and goodwill among the Arabs, particularly at the grassroots level. The NATO-US-Israel linkage is a problem that NATO policy-makers must confront in the NATO-Mediterranean dialogue.

In the Cold War period NATO officials were much more concerned with possible developments along the Central Front than the Mediterranean flank. However, NATO member states in the Southern Region were becoming more anxious about developments to their immediate south. One commentator could still assert in 1983, concerning the out-of-area issue and the Middle East, that "where the Soviet Union is not in evidence and Western interests are not deeply involved, there is a discernible and sensible tendency to keep clear and leave matters for regional organisations and local powers to sort out."[45] Nevertheless, one could perhaps argue, that the out-of-area issue led NATO officials to look increasingly at the Mediterranean through a north-south and south-south as well as an East-West perspective. In the post–Cold War period the Mediterranean remained much less a priority for NATO policy-makers in contrast to the Atlantic Alliance's relations with Russia and eastern Europe. NATO member states in the Southern Region, though, did successfully lobby for more attention to be given to the Mediterranean. The Atlantic Alliance's Mediterranean Initiative marked a new stage in the efforts of NATO to develop a more coherent policy toward territory that was traditionally referred to as out-of-area.

NOTES

[1]Ronald D. Asmus, Stephen Larrabee, and Ian O.Lesser, "Mediterranean Security: New Challenges, New Tasks," *NATO Review* 44, 3 (May 1996): 28–29.

[2]Jed F. Snyder, *Defending the Fringe—NATO, the Mediterranean, and the Persian Gulf* (Boulder, Colo.: Westview, 1987), 7.

[3]George Joffé, "The European Union and the Maghreb," in *Mediterranean Politics, Vol.1,* Richard Gillespie, ed. (London: Pinter, 1994), 23.

[4]Mark Stenhouse and Bruce George, *NATO and Mediterranean Security: The New Central Region* (London: London Defence Studies 22, Brassey's for The Centre for Defence Studies, 1994), 2.

[5]*British Foreign Office General Correspondence Documents (FCO),* FO371/67276 R 3783, March 14, 1947—Menemcioğlu Asks the Possibility of a Pact between Turkey, Greece, and Egypt.

[6]*FCO,* FO371/67276 R 11125, August 7, 1947—Erkin: A Mediterranean Security Pact.

[7]Don Cook, *Forging the Alliance: NATO 1945 to 1950* (London: Secker and Warburg, 1989), 204, 218: and, Douglas T. Stuart and William Tow, *The Limits of Alliance: NATO's Out-of-Area Problems since 1949* (Baltimore, Md., and London: Johns Hopkins University Press, 1990), 185.

[8]Douglas T. Stuart, "Continuity and Change in the Southern Region of the Atlantic Alliance," in *NATO at Forty: Change, Continuity and Prospects,* James R. Golden et al., eds. (Boulder, Colo., and London: Westview, 1989), 75.

[9]Stuart and Tow, *The Limits of Alliance,* 187–89.

[10]Yulağ Tekin Kukat, "Turkey's Entry to the North Atlantic Treaty Organization," *Diş Politika* (Ankara) 10, 3–4 (1983): 72–73.

[11]Roberto Aliboni, *European Security across the Mediterranean* (Paris: Chaillot Papers 2, Institute for Security Studies, WEU, 1991), 4.

[12]John Chipman, "NATO and the Security Problems of the Southern Region: From the Azores to Ardahan," in *NATO's Southern Allies: Internal and External Challenges,* John Chipman, ed. (New York and London: Routledge, 1988), 18.

[13]Maurizio Cremasco, "Italy: A New Role in the Mediterranean?" in ibid, 200.

[14]Joffé, "The European Union and the Maghreb," 23–24.

[15]NAA, Political Committee, Sub-Committee on the Southern Region, *Draft Interim Report,* AM 106 PC/SR (95) 1 by Mr. Rodrigo de Rato (Spain), Rapporteur (Brussels: International Secretariat, May 1995), 3.

[16]NAA, Presidential Task Force, *Final Report—America and Europe: The Future of NATO and the Transatlantic Relationship,* Co-Chairmen Loic Bouvard (France) and Charlie Rose (US) (Brussels: International Secretariat, September 1993), 21.

[17] Lawrence S. Kaplan, *NATO and the United States: The Enduring Alliance* (Boston: Twayne, 1988), 123.

[18] Ian O. Lesser, "Growth and Change in Southern Europe," in *Maelstrom,* 16.

[19] Fernando Rodrigo, "Southern European Countries and European Defense," in ibid., 152–53.

[20] John Chipman, "Allies in the Mediterranean: Legacy of Fragmentation," in *NATO's Southern Allies,* 83.

[21] Stuart, "Continuity and Change in the Southern Region of the Atlantic Alliance," 87.

[22] Maurizio Cremasco, "Do-it-Yourself: The National Approach to the Out-of-Area Question," in *The Atlantic Alliance and the Middle East,* Joseph I. Coffey and Gianni Bonvicini, eds. (Basingstoke and London: Macmillan, 1989), 147.

[23] For more details, see Gareth M. Winrow, "NATO and the Out-of-Area Issue: The Positions of Turkey and Italy," *Il Politico* (Pavia) 58, 4 (1993): 631–52; and Gareth M. Winrow, "NATO and Out-of-Area: A Post–Cold War Challenge," *European Security* 3, 4 (winter 1994): 617–38.

[24] *NATO Letter,* Vol.5, Special Supplement to no.1, January 1, 1957. Non-Military Cooperation in NATO—Text of the Report of the Committee of the Three, paragraph 32, 5.

[25] Meeting of the NAC, Paris, December 11–14, 1956, in *Texts of Final Communiqués 1949–74* (Brussels: NATO Information Service), paragraph 5, 101–102.

[26] Stuart and Tow, *The Limits of Alliance,* 211–16.

[27] The Harmel Report on The Future Tasks of the Alliance, issued by the Ministerial Sessions of the NAC, the Defence Planning Committee, and the Nuclear Planning Group, Brussels, December 13–14, 1967, in *Texts of Final Communiqués 1949–74,* paragraphs 14, 15, 201.

[28] Charles A. Kupchan, "Regional Security and the Out-of-Area Problem," in *Securing Europe's Future,* Stephen J. Flanagan and Fen Osler Hampson, eds. (London and Sydney: Croom Helm, 1986), 287.

[29] Kaplan, *NATO and the United States,* 157.

[30] Richard J. Payne, *The West European Allies, the Third World and US Foreign Policy: Post–Cold War Challenges* (New York, Westport, Ct., and London: Greenwood, 1991), 129.

[31] Document on Integrated NATO Defence issued at the Heads of State and Government Participating in the Meeting of the NAC, Bonn, June 10, 1982, in *Texts of Final Communiqués 1981–85, Vol.3* (Brussels: NATO Information Service), 77.

[32] Charles A. Kupchan, *The Persian Gulf and the West: The Dilemmas of Security* (Winchester, Mass.: Allen and Unwin, 1987), 207–08.

³³Ibid., 170.

³⁴Marjorie Lister, *The European Union and the South: Relations with Developing Countries* (London and New York: Routledge, 1997), 86.

³⁵Geoffrey Edwards, "Multilateral Coordination of Out-of-Area Activities," in *The Atlantic Alliance and the Middle East*, 238.

³⁶NAA, Civilian Affairs Committee, Sub-Committee on the Mediterranean Basin, *Report—Frameworks for Cooperation in the Mediterranean*, AM 259 CC/MB (95) 7 by Mr. Pedro Moya (Spain), Rapporteur (Brussels: International Secretariat, October 1995), 6.

³⁷Cremasco, "Italy: A New Role in the Mediterranean?" 226–27.

³⁸Christophe Carle, "France, the Mediterranean and Southern European Security," in *Southern European Security in the 1990s,* Roberto Aliboni, ed. (London, and New York: Pinter, 1992), 43.

³⁹For full details, see Sonia C. Cordenas, "The Contested Territories of Ceuta and Melilla," *Mediterranean Quarterly* 7, 1 (winter 1996): 118–31.

⁴⁰NAA, Political Committee, Sub-Committee on Out-of-Area Security Challenges to the Alliance, *Interim Report,* AC 182 PC/OA (85) 2 by Mr. Herrero de Minon (Spain), Rapporteur (Brussels: International Secretariat, October 1985), 21.

⁴¹*NATO to the Year 2000: Challenges for Coalition Deterrence and Defence,* Report of the Working Group of the Atlantic Council of the United States on the Future of NATO, Washington, D.C., March 1988. Chairman, Andrew J. Goodpaster, Rapporteur, Ian O. Lesser, 38.

⁴²Declaration of the Heads of State and Government Participating in the Meeting of the NAC, Bonn, June 10, 1982, in *Texts of Final Communiqués 1981–85, Vol.3,* paragraph 5e, 73.

⁴³NAA Political Committee, Herrero de Minon Report, 1985, 11.

⁴⁴Declaration of the Heads of State and Government Participating in the Meeting of the NAC, Brussels, May 29–30, 1989, in *Texts of Final Communiqués 1986–90, Vol.4* (Brussels: NATO Office of Information and Press), paragraph 32, 39.

⁴⁵Lawrence Freedman, "Conclusion," in *The Troubled Alliance: Atlantic Relations in the 1980s,* Lawrence Freedman, ed. (London: Heinemann, 1983), 161.

CHAPTER 4

The Southern Mediterranean and the Middle East in the Post–Cold War Era: Political Developments

INTRODUCTION

The aim of this chapter is to examine briefly the recent significant political developments in certain states in the southern Mediterranean and the Middle East that may have a bearing on the future of NATO's Mediterranean Initiative. Economic and social issues of note in these states will be analyzed in the next chapter. The focus will be on those non-NATO Mediterranean states participating in NATO's Mediterranean Initiative, excluding Israel and including Algeria—the latter not a member of NATO's Mediterranean dialogue.

Israel is not a typical Middle Eastern or southern Mediterranean state. In comparison with its Arab neighbors, Israel has very close relations with the West and with the United States in particular. Many Israelis would feel that their national culture had more in common with the national cultures of states of the northern rather than the southern Mediterranean. Israel is an economically advanced country. In 1998 its gross domestic product (GDP) per capita amounted to $16,138, ahead of the economies of Greece, Portugal, and Spain.[1] Compared to its neighbors, Israel has a developed, democratic, multiparty system of government. The Israeli armed forces are under civilian control. However, the Israeli political system also has its shortcomings. The electoral system, based on proportional representation, enables small, extremist parties to secure parliamentary seats by winning a small number of votes. These usually ultraorthodox religious parties can hold the balance of power in a coalition government. This could create problems for the Middle East peace process. Since its electoral success in 1996, the Netanyahu government has been dependent on the support of a handful of deputies from these parties. The problems in Israel regarding Palestinians have been discussed earlier.

Governments in the Maghreb and in Egypt, especially, are closely following events in Algeria. They are apprehensive about the possible seizure of power by radical Islamist elements in Algeria. A continued unstable Algeria could threaten to destabilize all of North Africa. If stability is restored and democratization encouraged, Algeria would likely be a leading candidate for inclusion in NATO's Mediterranean Initiative.

An understanding of the political, economic, and social issues and problems that the governments of these states might have to confront could help clarify the possible direction of NATO's Mediterranean dialogue. Serious difficulties at home could compel officials in particular Arab states to pay less attention and interest to the Mediterranean Initiative. Conversely, domestic troubles could prompt governments to attempt to upgrade their relations with institutions such as NATO, the EU, and the OSCE in order to bolster their position. Certainly, the government of each non-NATO Mediterranean state is faced with a different set of political, social, and economic problems that needs to be addressed. In these circumstances it is hard to envision how NATO officials will be able to ensure that the—in effect—separate dialogues conducted with the six non-NATO Mediterranean countries are parallel as much as possible.

The general impression in the West of states in North Africa and the Middle East is that they consist of authoritarian governments headed by powerful individual rulers and backed by politically and economically influential elite groups. Apparently, there is little or no democracy. Only certain political parties are tolerated. So-called religious parties are usually prohibited. There are close ties between the governing elites and leading military officers. Public opinion is ostensibly of marginal importance. There are few signs of an emerging civil society, and the prospects for democratization are not encouraging. How accurate is this picture? A closer look at individual countries will reveal that the political reality may not be quite so bleak in certain cases. There are signs and indicators of a gradual political and also economic liberalization occurring in parts of the southern Mediterranean and the Middle East. However, the fear among governing elites of the possible strengthening and spread of radical Islamism may stifle the movement toward increased democratization in certain states.

Islamism, in its general sense, may refer to groups aiming for a form of political and social organization that would use Islamic values as its basis. The term thus denotes a politicized form of Islam. There may be radical and moderate variants of Islamism. Movements based on radical Islamism are prepared to use revolutionary, if not violent means, to transform society. Moderate Islamist elements prefer to use peaceful, democratic means in order to achieve their goals.

RULERS AND ELITES

It has been noted that most of the individual rulers and elites in North Africa and in the Middle East tend to be more knowledgeable of, and more sympathetic to, the West than are their publics. (The one obvious and glaring exception is Libya's Gadhafi, who continues to spout anti-Western rhetoric.) These governing elites are concerned about a possible serious challenge to their authority from a popular movement from below, most probably inspired by radical Islamist forces. This fear, if not obsession, with societal security places obstacles in the path of democratization and creates a dilemma for governments in Europe and in the United States. NATO officials would not wish to be perceived as helping to shore up unpopular and illegitimate, but at the same time pro-Western, governments in the southern Mediterranean and in the Middle East. Yet, clearly NATO policy-makers would not be at ease if radical Islamist groups were swept to power in these states through free and open elections.

At present, throughout North Africa and the Middle East, states are governed to a large extent by a powerful president or hereditary king supported by various political, economic, and military officials and advisers. It has not proved possible to replace individual rulers by peaceful, democratic means. There are usually few constitutional constraints on the powers of these authoritarian heads of state. They are often able to appoint key ministers, rule by decree, and declare states of emergency.

Until his death in early 1999, King Hussein of Jordan had been actively involved in decision-making, particularly in matters affecting foreign policy and external security. King Hussein's successor, his son Abdullah, has the power to appoint the prime minister and consult with the premier on the choice of cabinet members. The king appoints all members to the senate or upper house of parliament. He and his family are greatly respected. Nevertheless, it seems that possible criticism directed toward the royal family is closely circumscribed. For example, in

May 1997 King Hussein had endorsed a restrictive press law that was supposedly aimed against sleaze. But the law prohibited any "news, views, and analyses" that disparaged the king or royal family, and also the armed forces and heads of friendly states.[2]

In Morocco there are indications that King Hassan is prepared to loosen a little his grip on power. The king has argued that his Islamic duties as the commander of the faithful or *Amir al-muminin*—a person of the highest religious authority, apparently genealogically descended from the Prophet—have prevented him from instituting a constitutional monarchy. Until September 1992 choosing the government in Morocco was a royal prerogative. After 1992 only the king's approval was required for the formation of a cabinet which should in theory reflect the composition of parliament.[3] In practice, the king is still able to reshuffle ministers and maintain a close watch on foreign and defense affairs especially. The interior minister, Driss Basri, is a particularly influential individual and a loyal associate of the king. The king has called for popular referenda in order to win public support for various policies. He is able to keep a keen eye over developments in towns and cities through a network of urban elites—*associations culturelles*. These are bodies created by the Royal Palace in order to oversee cultural and communal life. In the countryside rural elites owing allegiance to the king run municipalities and communes—*collectivités locales*. These bodies have a say over expenditure and also receive money from the state to provide local services. The king traditionally has been able to make use of and play off various patron-client groups through this framework of decentralized control.[4]

However, it seems that King Hassan has agreed to relax at least partially his firm control over parliament. A referendum held in September 1996 approved direct elections for the first time for all the members of a Chamber of Deputies in what was to be a new bicameral parliament. Previously, there had been a unicameral legislature with only two-thirds of its members directly elected and one-third indirectly elected from municipal bodies and professional associations over which the king had influence. General elections for the Chamber of Deputies were held in November 1997. The Chamber of Deputies has to cooperate with an indirectly elected upper chamber composed of what are mostly royalist appointees. This upper chamber has powers to amend laws and bring down the government of the day.[5]

In Tunisia, Ben Ali wrested the office of president from the ailing Habib Bourguiba in November 1987 and proceeded to consolidate his position as head of state. By the constitution the president is empowered to appoint the prime minister and other ministers. A former director of military security, Ben Ali has the backing of the police, the paramilitary national guard, the armed forces, and the Constitutional Democratic Rally (RCD) party, which dominates the legislature. The president nominates governors for each of the twenty-three provinces. Ben Ali was popularly elected president in April 1989 and March 1994. On both occasions he stood unopposed, and he secured as much as 99.91 percent of the vote in 1994.[6]

President Hosni Mubarak of Egypt is similarly powerful. The state of emergency, which has been in effect in Egypt since the assassination of President Anwar el-Sādāt by Islamic militants in 1981, is periodically extended by presidential decree. Emergency provisions include lengthy detention without trial, press censorship, and the trial of civilians by military courts. Mubarak has collected around himself a government of technocrats. Those critical of Mubarak accuse him of blocking political reform, and charge that officials close to the president are heavily corrupt.[7]

The power base of President Maaouya Ould Taya of Mauritania appears to be less secure in spite of the support of the armed forces. In the first popular multicandidate elections for president in January 1992, Ould Taya, who had been earlier installed as head of state by the military, obtained only 63 percent of the votes cast. His main challenger, and former president, Ahmed Ould Daddah, received 33 percent of the vote. In December 1997 Ould Taya was reelected president by a low voter turnout. The main opposition parties, citing previous alleged cases of electoral fraud, boycotted the election. There have been rumors of coup attempts against the Ould Taya presidency. For example, in October 1995 the Iraqi ambassador to Mauritania was expelled after being accused of supporting an attempted coup organized by the pro-Iraqi Arab nationalist Ba'athist movement based in Mauritania.[8]

The position of the Algerian president is open to question. In the popular presidential elections held in November 1995, General Liamine Zeroual beat off the challenge of three other candidates to win with 61.34 percent of the vote. Significantly, amendments to the constitution approved by a popular referendum in November 1996 stipulated that an individual could only serve as president for a maximum of two five-year terms of office. In Algeria, therefore, in contrast to other states in North Africa and the Middle East, one may not be head of state for life. But, in the meantime, the president still has effective control over parliament. In the newly established bicameral legislature, important pieces of legislation require the support of at least 75 percent of members of the upper chamber, the Council of the Nation. Here, in practice, the president's men may veto legislation, as one-third of the members of the upper chamber are presidential appointees. The president retains the right to rule by decree in exceptional circumstances. He still appoints all senior officials in the armed forces and in the judiciary, and also chooses the governor of the central bank and all governors of the administrative regions.[9] Nevertheless, the president is ultimately dependent on support from the armed forces. In September 1998 Zeroual, facing the threat of a general strike due to economic problems at home, and no longer apparently guaranteed the support of the Algerian military, declared that he would cut short his mandate and call for early elections to choose a new president in 1999. The preferred candidate of the Algerian military, former foreign minister Abdelaziz Bouteflika, was elected president in April 1999. This was only after the six other presidential candidates withdrew when it became clear that the military was about to fix the elections in Bouteflika's favor.

POLITICAL PARTIES

The authoritarian nature of government in the southern Mediterranean and in the Middle East has hampered the development of a genuine multiparty political system. In some instances political parties were not legalized until recently. Parties based on religion or ethnicity were invariably banned. Individual candidates can circumvent these restrictions by standing in elections as independents, although this impairs the prospects of a united opposition forming against the governing party. A party can continue to be based on religion and ethnicity in all but name by simply changing its official title. Or candidates from prohibited parties can stand for other parties that are sympathetic to their cause. Those parties closely associated with the president or king are usually the most successful. Opposition parties often complain of irregularities in the electoral process. Some parties have protested by boycotting elections or by refusing to accept cabinet posts. These protests have had little impact hitherto, although there are indications of a gradual, albeit slow, move toward the establishment of a more open and competitive electoral system that the international community in general could encourage.

In November 1993 Jordan held multiparty elections for the first time since 1956. National elections had been held earlier, in 1989. Those elections were called in order to pacify the public after riots had occurred following the introduction of economic reforms. Candidates could stand only as independents in 1989. Political parties were only legalized shortly before the 1993 elections. However, in order to undercut the increasing popularity of the Islamic Action Front (IAF), voters were compelled by the new electoral law, based on one man/one vote instead of proportional representation, to choose between party or tribal loyalties. The result was that a "tribal parliament" was elected. Forty-six of the eighty seats in parliament were secured by candidates with no party affiliation.[10] Nine opposition parties led by the IAF boycotted the national elections for the lower house of the National Assembly held in November 1997. The IAF had failed to persuade King Hussein to amend the electoral law for the 1997 election. According to the IAF, the electoral law was weighted in favor of rural constituencies where support for the king was strongest. As a result of the boycott, a parliament was elected largely composed of independents or tribal leaders who had campaigned on local issues.[11]

A party close to the president dominates the parliaments in Egypt, Tunisia, and Mauritania. The electoral process in Egypt is notorious for its ballot stuffing and voter intimidation. Only certain political parties are permitted to contest the elections. The governing party, the National Democratic Party (NDP) won 317 of the 444 seats available in the general election held in November–December 1995. The next most successful party, the liberal Wafd party, only obtained 6 seats. Independents won 114 seats. It was important for the NDP to achieve a substantial majority in the parliament. This left President Mubarak handily placed to secure the necessary two-thirds parliamentary vote to allow him to stand unopposed in the next presidential elections.

In Tunisia, the dominance of the presidential party, the RCD, is almost total. After Ben Ali assumed power in 1987, six other political parties were legalized but these have proven no match for the RCD. When multiparty elections were held in April 1989, the RCD swept the board and took all of the seats in the Chamber of Deputies. In order to prevent a repetition of the 1989 results, in the next general election held in 1994, nineteen seats in parliament were reserved for four opposition parties. These opposition parties, between them, could only muster 2.87 percent of the national vote. In the municipal elections held the following year, the RCD won 4,084 out of the 4,090 positions that were contested. It seems that considerable pressure is put on the public to vote in such large numbers for the RCD.[12] In what appears to have been a token gesture, in

1995, opposition deputies were allowed to sit on parliamentary commissions.

In Mauritania, in July 1991, the military-controlled government legalized political parties. This was after support was obtained through a popular referendum. The party closely connected with the president, the Republican Democratic and Social Party (PRDS) triumphed in the general elections of 1992 and 1996, taking sixty-seven and then seventy of the seventy-nine seats that were contested. The main opposition party, the Union of Democratic Forces (UDF) led by Ould Daddah, boycotted the elections in 1992 after complaining that the results were going to be rigged. In 1996 the UDF pulled out of the second round of voting after protesting that the first round had been marred by electoral fraud.

King Hassan of Morocco has based his authority more on the support of various urban and rural elite groups who controlled the indirectly elected parts of parliament, and relied much less on the backing of a particular political party. As a consequence, in recent years, governments of nonparty technocrats have been formed. In the general election of June 1993, two large parties traditionally critical of the king—the Socialist Union of Popular Forces (USFP) and the Istiqlal Party (PI)—formed an electoral alliance and secured many votes. The king offered the two parties thirty-one of the thirty-five portfolios in the government. These were refused on the grounds that the most important posts were not made available. The leaders of the USFP and the PI also complained that the results in the indirect elections (for one-third of the seats in the then single-chamber parliament) were fixed in favor of the more pro-royalist center-right Constitutional Union (UC). The USFP and PI leaders argued, because of the success of their parties in the direct elections, that the authorities had manipulated the results in the indirect elections.[13] National elections to the lower house of parliament were again held in November 1997. A center-left coalition group headed by the USFP and the PI would finally agree to take office in March 1998. Center-left parties had won 102 of the 325 seats. The king chose as prime minister a member of the center-left coalition group. But the new government must rely on the continued support of a centrist party headed by the king's brother-in-law in order to maintain a majority in the lower house. The king also insisted on keeping on five ministers from the previous government, including Interior Minister Basri.[14] Apparently, Basri remains the most powerful minister. It was he who issued a controversial order for riot police to charge at a peaceful rally of jobless graduates outside the parliament in Rabat in October 1998.[15]

Algeria had been an example of a one-party state under the National Liberation Front (FLN) until a serious economic crisis in 1988 forced the ruling authorities to open up the political system and legalize other parties. Even an Islamist party, the Islamic Salvation Front (FIS) was allowed to operate. As is well known, the FIS was about to triumph in the general elections in January 1992 when the military intervened, canceled the elections, called a state of emergency, and outlawed the FIS. A general election would not be held until June 1997. Parties based on religion or ethnicity were in theory banned, but in practice moderate Islamist groups and parties representing the Berber minority were allowed to participate in the general election in 1997. Not surprisingly, a party that had recently been formed by the president and his military supporters, the National Democratic Rally (RND), secured the most votes. Opposition leaders complained of voting irregularities. UN observers declared that they had encountered problems in monitoring the polls and stated that the security forces had been conducting checks on how people had voted. Nevertheless, the RND did not trounce the opposition. Of the 380 seats in the National Assembly, the RND won 155. The moderate Islamist Movement for a Peaceful Society (MPS) won 69 seats; the FLN 64; another moderate Islamist party, En Nahda 34; and the socialist Berber party (the Front of Socialist Forces [FFS]) and the pro-Berber and anti-Islamist Rally for Culture and Democracy (RCD) both won 19 seats. The subsequent government was dominated by the RND, but in a positive step toward further democracy, the FLN and MPS were offered and accepted seven posts each in the new cabinet. The electoral success of the RND was again in evidence in the local elections of October 1997 when the party secured 7,242 of the 13,123 positions vacant. The next most successful party, the FLN, only obtained 2,864 seats on municipal councils. Again, opposition party leaders complained of electoral malpractice.

RELIGIOUS RADICALISM

Western concerns about the impact of the possible spread of radical politicized Islam or radical Islamism in North Africa and the Middle East have been noted earlier. There are fears that further growth of religious radicalism could turn the Mediterranean into a highly unstable area. In these circumstances, some reports refer to the likelihood of a large-scale wave of migration from the southern Mediterranean to countries such as Spain and Italy. Many in the West associate radical Islamism with violence and terrorism. Certainly, at the time of writing, atrocities were being committed in Algeria. Extremist Islamist groups are largely to blame. These groups were responsible for a series of bomb attacks in Paris and for the hijacking of the Air France Airbus in Algiers in December 1994. Innocent civilians have also been the victims of terrorist attacks by (Islamist) Hamas in Israel and by the (Islamist) Jama'at Islamiyya and other more shadowy groups in Egypt. It is highly unlikely that Iran is coordinating many of these activities, through the peoples and governments of predominantly Sunni Muslim and Arab North Africa are traditionally suspicious of the Shi'ism of Persian Iran.[16]

There is also alarm among some circles in the West at the possibility that the Muslim population in Europe may become radicalized as a result of future developments across the Mediterranean.[17] This scenario is exploited—and overexaggerated—by extreme right-wing and anti-immigrant groups, especially in France.

Some Western commentators have suggested that in order to check the rise of radical Islamism in North Africa and the Middle East, moderate Islamist groups should be allowed to participate in elections. According to this argument, the democratic process in the states in question would be strengthened by the inclusion of such groups. These moderate Islamist elements would have to agree to abide by the principles of pluralism and democratic change. This would hopefully quell the fears of those who believe that the Islamists—including moderate Islamist groups—actually espouse the principle of "one man, one vote, one time." There is an assumption that the Islamists would triumph in the first genuinely open and competitive elections and then refuse to hold further elections. There is also the general recognition that the serious economic and social problems in the southern Mediterranean and in the Middle East need to be successfully tackled in order to diminish the appeal of radical, politicized Islam.[18] Unemployed and disillusioned urban youth are particularly attracted to radical Islamism. But how will Western officials be able to convince largely authoritarian governments that they should tolerate, if not encourage, moderate Islamist parties when these same governments are fearful of their publics' interest in any form of politicized Islam? And would these governments also be prepared to undertake serious economic reforms that could damage the interests of those progovernment bureaucrats and other elite groups who have benefited financially from the lack of proper reforms hitherto?

At a seminar in Paris in June 1996 organized by the WEU Institute for Security Studies, attention was drawn to the different approaches taken to Islamic political movements by different states in North Africa and the Middle East. One group, labeled the "eradicators," has sought to totally exclude radical Islamists from political life. This group was apparently active in Algeria and, with subtle differences, in Egypt and Tunisia. Others were more in favor of "assimilation," attempting to include moderate Islamists in political life but not allowing them to participate in elections. This was reportedly happening in Morocco. An integration model was also noted, in which an effort would be made to completely integrate Islamist forces—presumably moderate ones—into political life. It was argued that this policy was previously employed to varying degrees and then abandoned by governments in Algeria, Egypt, and Tunisia.[19]

Governments in North Africa have anxiously followed developments in Algeria.[20] The serious and bloody unrest that has afflicted Algerian society since the cancellation of the second round of elections in 1992 is one of the principal reasons why Algeria has not been included in NATO's Mediterranean Initiative. The FIS had swept local and provincial elections in 1990 and had been poised for another victory in 1992 because of public discontent with government corruption, economic mismanagement, soaring inflation, and rising unemployment. The military intervention terminated this attempt to integrate Islamist elements into the Algerian political arena. In the escalating violence that ensued, the Islamic Salvation Army (AIS) of the FIS clashed with the military authorities. Brutal terrorist operations were mounted by the more extremist Armed Islamic Group (GIA). The "eradicators" within the military prevailed upon President Zeroual (originally appointed president in January 1994) to continue to exclude the FIS from the political process. However, there have been recent attempts to assimilate moderate Islamist groups. Sheikh Mahfoud Nahnah, the leader of the nonviolent Hamas—later renamed the MPS—came second behind Zeroual in the presidential elections in November 1995, winning a quarter of the vote. And, as previously noted, the MPS and another moderate Islamist group, En Nahda, participated in the general and local elections in 1997, with the MPS even becoming a partner in a three-party national-coalition government. The central authorities remain highly suspicious of the FIS. In mid-July 1997 the FIS leader Abassi Madani, who had been sentenced to twelve years imprisonment in 1992, was released on parole on the condition that he not become involved in politics. By the start of September

1997 Madani was back under house arrest after writing a letter to UN Secretary-General Kofi Annan calling for an immediate halt to the bloodshed in Algeria.

Partly in reaction to events in Algeria, the government in Tunis has clamped down hard on Islamist groupings in Tunisia. There would seem to have been no serious attempt to integrate such groupings into Tunisian political life in recent years. In 1986–87 then President Bourguiba had struggled to contain the increasingly powerful Islamic Tendency Movement (MTI), which, with student support, had been waging a campaign of civil disobedience.[21] Ben Ali closed down the MTI and refused to grant political-party status to the more moderate En Nahda. Although they failed to win any seats, candidates sympathetic to En Nahda received 15 percent of the total vote in the 1989 national elections. Ben Ali must have been especially alarmed at the electoral performance of pro–En Nahda candidates in the capital where they obtained 30 percent of the vote. Economic problems and continued student activity appear to account for these results. Many of En Nahda's leaders were arrested in 1991 after a crowd attacked the headquarters of the RCD at the time of the Gulf War in February 1991. Rachid Ghannouchi, the leader of En Nahda, was granted asylum in London in 1993. It would appear that the Ben Ali government has made no attempt to differentiate between En Nahda and the much more radical Tunisian Islamic Front.[22]

Ben Ali is clearly apprehensive about a possible spillover from the turmoil in Algeria. In August 1994 there were reports that a Tunisian version of the FIS had been formed. Its leader, Mohammed Aliel-Horani was advocating the need for an armed uprising in Tunisia.[23] In February 1995 the GIA claimed responsibility for an attack against the Tunisian border town of Tamerza. There is certainly close cooperation between the intelligence and security services of Algeria and Tunisia.

In comparison with Tunisia, the central authorities in Egypt have been more tolerant of the activities of moderate Islamist elements. There has been a clamp-down on more radical Islamist groups, which was to be expected given the circumstances behind the assassination of President el-Sādāt in 1981. Members of el-Jihad and Jama'at Islamiyya, which have been involved in acts of violence and terrorism against official and tourist targets, have been arrested. Prior to the massacre of over sixty foreign tourists at Luxor in November 1997, the government believed that it had obtained the upper hand against Islamist extremists. It has become customary for the central authorities to turn a blind eye on the activities of the nonviolent Muslim Brotherhood, although that group has not been able to establish its own political party. Its members have stood as independents or run under the banner of another party in parliamentary elections. In a change of policy, the government closed down the headquarters of the Muslim Brotherhood in Cairo immediately before the 1995 general election. It seems that the authorities have become concerned at the Muslim Brotherhood's successful infiltration of professional unions and syndicates in recent years. These associations of doctors, lawyers, engineers, and the like have come to constitute a major element in an emerging Egyptian civil society. In 1993 and 1994 the government felt compelled to interfere in the voting procedures of these syndicates in order to prevent the election of more Islamist candidates.[24]

In spite of the restrictions imposed on the activities of the Muslim Brotherhood, in order to meet the Islamist challenge, the Egyptian government has allowed more religious programs to be aired on television and has permitted the formation of private voluntary organizations based on Islam. The religious scholars (*ulema*) of the Al-Azhar University have been given a leading role in promoting Islamic principles and an Islamic way of life in Egyptian society. As a consequence, artistic and intellectual freedoms have been threatened.[25] The stance of the Egyptian government toward Islam is a subject of debate. Did the central authorities initially attempt to integrate or assimilate the Muslim Brotherhood and fail? And at present, with regard to the government and Islamist groups in society, it is not clear who is endeavoring to integrate or assimilate whom.

Morocco may be regarded as an example of a state that has succeeded in assimilating moderate Islamist groups without allowing Islamist political parties to participate in elections. Thus, the Justice and Charity association has been able to function as a charitable body. The Movement of Reform and Renewal in Morocco (HATM)—which gets its support from intellectuals and the middle class, and advocates peaceful, democratic means to achieve its goals—has been able to enter politics indirectly through affiliating itself with a legal opposition party. More radical Islamist groups are less influential, although they were able to organize large pro-Iraq demonstrations in early 1991. And in February 1994 Islamist students clashed with their leftist counterparts.[26] Apparently, though, unlike in Algeria or Egypt, violent Islamist groups pose no serious challenge to the king's authority. This is because of the status of the king as commander of the faithful. Thus, in contrast to Tunisia, the king appears more prepared to tolerate peaceful manifestations of politicized Islam.

It would seem that the government of Mauritania has also sought to assimilate moderate Islamists even though Islamist parties are prohibited by the constitution. In 1990 the Islamic Movement of Mauritania (HASM) was founded. This group aimed to overthrow the government by violent means if necessary. It had developed contacts with radical Islamist groups in Tunisia and Sudan. Probably fearful of the potential for increased support for such radical groups, in 1994 and 1995 the government called on the army to ruthlessly suppress urban riots. Evidently, Islamist radicals may be found in the opposition UDF and even in the ranks of the governing PRDS.[27] Islam does play an important role in Mauritanian society. The state is even officially titled the Islamic Republic of Mauritania. However, apparently, this official designation is aimed at cementing the allegiance of various tribal groups in what is largely a communally segmented society.[28] In practice, the ruling authorities, it seems, are only prepared to tolerate an official and nonpolitical form of Islam.

In Jordan, did the late King Hussein attempt to control Islamist forces by seeking to integrate them fully into Jordanian society? Traditionally, there have been close links between the Hashemite monarchy and the moderate Muslim Brotherhood in Jordan, in order to counter the appeal of the Pan-Arab left and the radical Islamic right. But with the advent of multiparty politics in Jordan, politicized Islamic groups have become increasingly critical of government corruption. The moderate and nonviolent IAF, closely associated with the Muslim Brotherhood in Jordan, was able to win thirty-two seats in the 1989 elections with its candidates running as independents. In spite of the tinkering of the electoral process in 1993, the IAF still succeeded in acquiring sixteen seats and became the largest opposition faction in parliament.

The IAF has performed an important role in Jordanian politics. It has steadfastly opposed the peace treaty Jordan signed with Israel in 1994. The IAF gives those Palestinians living in Jordan who are critical of the Middle East peace process the opportunity to take part in politics, without being labeled Palestinians.[29]

In spite of this accommodation of moderate Islamist forces, it would seem, though, that in order to advance further the process of democratization in Jordan, various obstacles must still be overcome. For example, the IAF performed surprisingly poorly in local elections held in 1995, losing even the traditional Islamist stronghold of Zarqa, Jordan's second largest city. The IAF, perhaps with some justification, protested that the voting was rigged. Democracy suffered a more serious setback in

Jordan in late October 1997, only a week prior to the general election, when the IAF announced that it was pulling out of the elections. IAF leaders declared that they were protesting what they believed to be the gradual erosion of parliament's independence by a government that had tightened restrictions on the press, had committed itself to peace with Israel, and had introduced painful economic reforms. As previously noted, IAF leaders were also especially upset at their failure to revise the electoral law that they believed discriminated against them.

Governments in North Africa and in the Middle East have resorted to different methods in order to deal with Islamist elements in their societies. Repression of these groups is clearly evident in Tunisia, and to a lesser degree in Algeria. The ruling authorities in Jordan, Mauritania, and Egypt have pursued their own particular policies to come to terms with varying forms of moderate politicized Islam. Morocco appears to be a special case because of the unique role of the king. For the foreseeable future no exclusively Islamist government is likely to take office in the southern Mediterranean, although political Islam or Islamism remains a powerful latent force and is arguably the most popular movement in the area.[30]

How the West may be able to encourage the further accommodation of moderate Islamist groups in political life in the southern Mediterranean and in the Middle East is not clear. What is evident is that NATO is conducting a dialogue with various Mediterranean states that have adopted a whole range of different policies toward their own particular Islamist groups. The overt repression of politicized Islam in its various guises by the Ben Ali government has not prevented Tunisia's inclusion in NATO's Mediterranean Initiative. It is important that policy-makers in the West be aware of the manifold forms of politicized Islam and not automatically label all supporters of Islam violent, anti-Western, radical extremists. It will be seen that public references to a simplified and distorted image of Islam resulted in serious difficulties for NATO officials at the very outset of the Atlantic Alliance's dialogue with the Mediterranean.

THE ROLE OF THE MILITARY

The topic of military-civilian relations is one that is being discussed by NATO officials with representatives from North Africa and the Middle East as part of NATO's Mediterranean Initiative. Policy-makers in Brussels, therefore, should have a proper appreciation of the role of the armed forces in political life in the southern and eastern Mediterranean. In some states, the military regards itself as a bulwark against the spread of religious radicalism. The armed forces in Turkey, a NATO member state, also believe that it is their duty to perform such a role. But in some instances, in their support for authoritarian ruling elites, the armed forces may also be hampering the development of democracy.

The military is most obviously and directly involved in political life in Algeria. It prevented the FIS from taking office in 1992. The "eradicators" within the Algerian armed forces prevented President Zeroual from possibly embarking on a dialogue with the FIS. The military is eager to maintain its grip on political power because of the intention of the FIS to adhere to the provisions of the 1989 constitution. These provisions refer to the removal of the armed forces from political life.[31] The Algerian opposition parties—including the FIS—that signed the Rome Contract in January 1995 reiterated their demand that the armed forces should withdraw from politics in accordance with the constitution. However, one should also bear in mind that the armed forces have been at the center of every government since Algeria won its independence in 1962. The army has served as a shield behind which weak civilian rulers have been able to govern. Following the line of this argument, the armed forces in Algeria should not be regarded as a "usurper of political authority."[32]

In Mauritania, arguably, the Ould Taya government is still ultimately dependent on the military. President Ould Taya himself holds the rank of colonel. The military had previously overthrown the Ould Daddah government in 1978. In addition to combating civil unrest in the shanty towns in 1994 and 1995, the military had also waged, for a number of years until the early 1990s, a campaign against units of the Mauritanian African Liberation Forces (FLAM), which had launched operations from Senegalese territory.[33]

The armed forces have also played a significant role in Moroccan politics. In the past the armed forces were a serious internal threat to the monarchy in Morocco. There were attempted military coups in 1971 and 1972 and one may also have been planned in 1983. The conflict over the Western Sahara enabled the king to ensure that the army would have less time to become involved in domestic politics. The image of the armed forces in Morocco has also improved as a result of the army's operations in the Western Sahara.[34] It seems that King Hassan has secured a firm control over his military officers in recent years.

The army became an influential actor in Egyptian politics after leading military officers carried out a coup in 1952. The reputation of Egypt's armed forces was tarnished after the humiliating defeat in 1967 at the hands of the Israelis. The military thereafter played a less active role in Egyptian politics. However, President Mubarak still needs the backing of his military in order to contend with Islamic-sponsored terrorism. The president must thus be anxious at the spread of radical Islamism among the poorly paid ranks of the armed forces. Extremist Islamist groups are also apparently attempting to infiltrate the lower ranks of the security services, which receive even less pay than their military counterparts.[35]

In Tunisia, the military played a leading role in supporting Ben Ali when he came to power. Tunisian army officers share the president's dislike for any politicized form of Islam. The armed forces in Tunisia are the traditional custodians of law and order. They effectively crushed public riots in 1978 and 1984.[36] Ben Ali has the firm backing of his armed forces. Likewise, King Abdullah of Jordan, who is Supreme Commander of the Armed Forces, is able to depend on the loyalty of the Jordanian military. His father preempted a possible military coup as far back as April 1957. The army was then reorganized. The number of Palestinians—who were regarded as politically suspect—within the armed forces was reduced and there was an increase in the recruitment of troops from local tribes.[37]

PROSPECTS FOR DEMOCRATIZATION

NATO, the OSCE, the Mediterranean Forum, and the EU in its Barcelona Process, have all expressed a keen interest in the fostering of a civil society in the southern Mediterranean and in the Middle East. Hopes and expectations have been raised with the decline in the appeal of the one-party state following the end of the Cold War and the demise of the Soviet Union. However, in general, the view still prevails in the West that there is little evidence of the emergence of a genuine civil society in the Middle East. This chapter has already drawn attention to the power and authority wielded by authoritarian leaders in societies where there are often limits placed on multiparty politics, where even the activities of the most moderate Islamist groupings may be strictly curbed, and where the military remains as a force in the background that may intervene in the political arena at any moment in order to guarantee its perceived interests.

This is not the place to enter into the debate about whether Islam is compatible or not with democracy—concerning the argument whether sovereignty should be invested in God or in the people. One may contend that most people, if given the opportunity, would prefer that their rulers be held accountable for their actions and accountability; this is a key element of democracy. It remains to be seen whether moderate Islamist parties would act as good democrats if they were voted into political office rather than attempting to apply the principle "one man, one vote, one time." No Islamist party has come near to electoral victory in the six non-NATO Mediterranean states that are participants in the Atlantic Alliance's Mediterranean Initiative. The religious Welfare party in Turkey was touted as an example of an Islamist grouping that was willing to follow the rules of democracy but even this party, when heading government for one year, had to share power with another secular party. Under pressure from the Turkish military, the Welfare party was forced to relinquish control of the government. The party was then banned because it had apparently encouraged actions that were hostile to the secular nature of the Turkish state. At the time of writing, though, many deputies in the Turkish parliament who had formerly been members of the Welfare party had simply formed another religious party known as the Virtue party.

As noted in an earlier chapter, the democratization of a traditionally nondemocratic society may result in, at least in the short and medium term, heightened tensions, as governing elites are likely to be reluctant to share or hand over power to different social groups. Unlike the past democratization of western Europe, governments today in North Africa and in the Middle East face the dual problem of pressures for democratization, on the one hand, and the need to complete state-building on the other. Democratization should complement rather than contradict state-building. Incumbent ruling elites should find means to peacefully accommodate other political groups and give these groups more access to political and economic power. Previously exploited ethnic minorities and moderate Islamist groupings, for example, could presumably be accommodated within such an arrangement. In the transition from an authoritarian to a democratic society in North Africa and the Middle East, the current ruling elite could share power with the secular opposition and also with moderate Islamists through free and open elections and equal access to the mass media.[38]

The serious economic problems governments are confronted with in the southern and eastern Mediterranean will be examined in the next chapter. Extensive restructuring of the economies of these states is essential. An independent judiciary and the proper rule of law are imperative in order to obtain a respect for private property rights and encourage foreign investment. Unpopular but necessary austerity measures prepared by the International Monetary Fund (IMF) must be implemented although these would lead to cuts in subsidies, increases in taxes, and more unemployment for a period. In times of economic hardship it is wise for governments to embark on programs of political liberalization in order to win over popular support and give people the sense that they are sharing in the making and execution of painful decisions.[39]

The prospects for democratization in North Africa and the Middle East, therefore, will be enhanced if there is an improvement in the economic performance of states that, in turn, ought to curb the appeals of extremism and radicalism. Moderate Islamist parties should be allowed to operate. Ruling elites should be more tolerant of opposition groups, in general, provided they espouse peaceful, democratic causes. Peaceful relations between states would also provide a more favorable environment for democratization. The importance of a durable and lasting Palestinian-Israeli and Arab-Israeli peace settlement can not be overstated.

Perhaps somewhat optimistically, Augustus Richard Norton has taken the line that a civil society is already reasonably well developed in the Middle East. He has referred to the positive impacts in the Middle East of the rapid political liberalization in central and eastern Europe and in parts of the former Soviet Union. Norton even regarded the popular protests against the Gulf War in 1991 in a favorable light, noting that public opinion was becoming increasingly important in the Arab world. According to Norton, many Islamist groups are seeking to work within the current political system. NGOs in general are continuing to proliferate with the spread of women's associations, business lobbies, human rights organizations, journalists groups, and so on. In Norton's opinion, civil society is a necessary though not sufficient condition for the advancement of democracy. He cautions that governments will still need to play a controlling and intermediary role vis-à-vis a developing civil society, otherwise there could be disorder and chaos.[40]

One should also note, though, that NGOs in North Africa are often political actors with close links to different sections and factions within a state. These NGOs may have ties to various party political projects.[41] In practice, therefore, it is not always a simple case of state versus civil society, as it were, nor an example of a civil society totally independent from a state. Furthermore, one should not necessarily assume that an emerging business middle class benefiting from economic reforms would automatically be a force in favor of political liberalization and democracy. Business lobbies, on occasion, may be more interested in profits than in democracy.

Clearly, in any discussion of civil society one must not disregard the role of the masses. Throughout North Africa and the Middle East in recent years there have been disorganized and spontaneous popular revolts against the economic policies of the governments. This was the case in Morocco in 1981, 1984, and 1990–91; in Algeria in 1986 and 1988; in Tunisia in 1984; in Egypt in 1986; and in Jordan in 1984, 1989, and 1996. In 1991, at the time of the Gulf War, anti-Western and pro-Iraqi demonstrations were held in Algeria, Mauritania, Morocco, and Tunisia. Pro-Iraqi rallies were also held in Jordan in February 1998 when the United States and Britain were on the verge of launching air strikes against the Baghdad government because of obstacles placed in the way of the work of the UN arms-inspection teams in Iraq. Certainly, then, it is not the case that the masses in the Middle East have always been passive and subservient. There has been a marked increase in anti-Western sentiment among the general public in the aftermath of the Gulf War and the crisis in Bosnia. The masses believe that their economic and social problems are due to the bankruptcy of their governments and the disinterest and lack of concern on the part of the West. It will not be easy for Western governments to win over public confidence in the Middle East.

Tim Niblock has differentiated between the governments of those states in North Africa and in the Middle East that have sought to introduce a full parliamentary democracy and those that have aimed to achieve only a "managed democracy." According to Niblock, until 1992 the central authorities in Algeria were aspiring toward the realization of a genuine parliamentary democracy. He believes that the government in Jordan is still pursuing this goal. Egypt, Morocco, and Tunisia are listed as examples of "managed democracy." Niblock argues that in this second category of states, real power remains in the hands of a small ruling elite. There is an electoral framework of a sort but restrictions are placed on certain political parties, including those that are probably the most popular.[42] Therefore, it would seem that in so-called managed democracies there are actually serious curbs on political activity.

The internal upheavals in Algeria that seriously set back democratization in that country did not only check the growth of politicized Islam. In recent years the three million Berber minority has also protested against what they perceive to be discriminatory treatment. One positive indicator here, though, is that the central authorities have not prevented the FFS and the RCD from attempting to mobilize the Berber vote. Some Berbers continue to press for autonomy for the largely Berber-populated Kabylia region. Arabic remains the sole national and official language. In November 1996 a general strike was called in Berber-populated areas against the proposed constitution, which did not give official status to the Berber language. In summer 1998 there was further unrest after the assassination by Islamic militants of Lounes Matoub, a popular Berber singer who had just returned from exile in France. Government-organized anniversary celebrations of Algerian independence also provoked Berber protests. In addition to economic difficulties and the rise of radicalized Islam, the authorities in Algeria, therefore, also have to confront a problem of ethnic division. State- and nation-building in Algeria would appear to be far from complete. In Morocco, by contrast, the unofficial recognition by the central authorities of the Berber minority—which is larger in number than the Berber population in Algeria—has forestalled any potential conflict.[43]

Is Jordan a full-fledged democracy? The extensive powers of the king, and the complaints expressed by Islamist groups, appear to suggest that Jordan may be more accurately labeled an example of a managed democracy. It remains to be seen whether King Abdullah will decide to seek a new accommodation with the Islamists in Jordan. Although the president remains dominant in Egypt, and a genuine multiparty system has yet to form, the mushrooming of various NGOs is a sign of democratization. In Mauritania, the president and the military seem to be prepared to tolerate only limited opposition. As in the case of Algeria, ethnic cleavages in society pose added problems. For example, in the March 1992 national elections, the ruling party mobilized the vote of the Arab and Arabized Mauritanians in the rural and northern areas of the country. The opposition sought to win votes from the black Mauritanians in the capital and in the south of the country.[44] The Ben Ali government in Tunisia has been criticized by many in the West for its abuses of human rights. There are reports of widespread torture and restrictions on the freedoms of speech and press. There is no independent judiciary. However, the rights of women in Tunisia are respected, and Tunisian women are probably the most emancipated in the Arab world.[45]

Morocco may become an important test case for the prospects for further democratization among Arab states. The king may choose to relax his grip over Moroccan society safe in the knowledge that radical and politicized Islamic opposition groups are unlikely to pose a threat because of his widely acknowledged religious status. Perhaps the successor of King Hassan will opt to perform a role more suited to a constitutional monarch. There are also signs that the electoral system is becoming more open. It even seems that some of the younger and more educated elite groups are becoming more politically conscious and more in favor of further democratization. It has been suggested that these groups are pressing for the monarchy to move beyond a "scriptural democracy" toward a "substantive democracy."[46] However, it would seem that the hardline Interior Minister Basri must be removed before real political liberalization may develop in Morocco.

King Hassan cannot afford to overlook the demands of his people. The king was clearly out of touch with popular feeling when he dispatched a Moroccan troop contingent to the Gulf in 1991 on the side of the international coalition against Saddam Hussein. In the face of riots, demonstrations, and a general strike, the king was compelled to announce that troops were sent to the Gulf not to liberate Kuwait but to defend Saudi Arabia and the holy sites there.[47]

There could also be other complications with regard to the prospects for democratization in Morocco. The king may have to contain popular pressure for a Moroccan takeover of the enclaves of Ceuta and Melilla. And he will have to ensure the continued support of his public and also the armed forces over the question of the Western Sahara, although this clashes with the national aspirations of the Sahrawi people and may also result in a future conflict with Algeria.

The future of democratization in Morocco is also tied in with the question of succession to the throne. A succession crisis is possible. Religious officials would choose a successor from among the royal princes. The eldest son would not automatically become the next king.[48] A choice of successor could be controversial, particularly if the new king would not be prepared to venture further along the path of constitutional reform.

CONCLUSION

Officials in the West should be mindful of the problem of "conditionality." For instance, they may insist that certain conditions such as political reform be met before Western economic aid be extended to a particular country. Economic sanctions may even be threatened in an effort to encourage democratization. Arab public opinion would most probably react vehemently by claiming that the West was seeking to impose its values on their societies.[49] With reference to democracy and the Middle East, the West has already been accused of double standards. For example, Arab leaders have criticized the West for sheltering certain exiled religious leaders who are purportedly opposed to democracy. As previously noted, Rachid Ghannouchi, the leader of the En Nahda party in Tunisia, is in exile in London. A number of FIS officials have fled from Algeria and are active in western Europe. The nondemocratic but strategically important Saudi Arabia has not been pressured by Western governments to liberalize its closed political system. Concerning NATO's Mediterranean Initiative, however, it does not appear that certain political conditions need to be fulfilled before states will be admitted to the NATO-Mediterranean dialogue. There are serious shortcomings with regard to the lack of democratic freedoms in most of the non-NATO states in question. Other considerations, including strategic issues, obviously come into play. It seems, though, that Algeria is not going to be admitted to the NATO-Mediterranean dialogue until internal stability is restored and further political liberalization encouraged. The non-NATO states currently included in the dialogue do not have to contend with the problem of serious turmoil at home.

There are signs of a possible, gradual process of democratization in some states in the Middle East and in North Africa. The end of the Cold War has encouraged this process. Previously, several governments were impressed by the Soviet model of a closed one-party state based on a command economy. Clearly, the process of democratization is still in its infancy in most instances. Western governments should not attempt to interfere in the internal affairs of Arab states in the name of democracy, as this would most probably only result in a popular backlash against the West. This is an issue that is also connected with security perceptions and how peoples, especially in the southern Mediterranean, might perceive states to their north as a threat. This issue will be analyzed in the next chapter. How Western officials might help improve the economic and social conditions in North Africa and in the Middle East, without necessarily attaching political strings to offers of aid and assistance, is also discussed.

NOTES

[1] See *The World in 1999* (London: The Economist Group, 1998).

[2] "Jordan: A Not So Loyal Opposition," *The Economist*, August 30, 1997.

[3] George Joffé, "Elections and Reform in Morocco," in *Mediterranean Politics, Vol. 1,* Richard Gillespie, ed. (London: Pinter, 1994), 213, 212.

[4] George Joffé, "Democracy in the Maghreb," in *The Middle East in the New World Order,* 2d ed., Haifaa A. Jawad, ed. (Basingstoke and London: Macmillan, 1997), 53–55.

[5] "Morocco: On the Road to Something New," *The Economist*, September 28, 1996.

[6] NAA, Civilian Affairs Committee, Sub-Committee on the Mediterranean Basin/Sub-Committee on the Southern Region, *Report—Tunis, Tunisia, 6–8 June 1995*, AM179 rev.1 CC/MB (95) 6 (Brussels: International Secretariat, September 1995), 2–3.

[7] Cassandra, "The Impending Crisis in Egypt," *Middle East Journal* 49, 1 (winter 1995): 18–19.

[8] *Keesings Record of World Events 1995*, 40758; and Anthony G. Pazzanita, "Political Transition in Mauritania: Problems and Prospects," *Middle East Journal* 53, 1 (winter 1999): 44–58.

[9] Hugh Roberts, "Algeria: A Controversial Constitution," *Mediterranean Politics* 2, 1 (summer 1997): 188–92.

[10] Glenn E. Robinson, "Can Islamists Be Democrats? The Case of Jordan," *Middle East Journal* 51, 3 (summer 1997): 376.

[11] "Jordan: Drifting Apart," *The Economist*, November 8, 1997; and Raad Alkadiri, "Profile: Jordan's Fading Democratic Façade," *Mediterranean Politics* 3, 1 (summer 1998): 170–72.

[12] NAA, Civilian Affairs Committee, Tunisia Report, 1995, 3–4.

[13] Joffé, "Elections and Reform in Morocco," 214–16.

[14] "Morocco: The King's Gift to His Country," *The Economist*, March 21, 1998.

[15] "Morocco: Back to the Batons," *The Economist*, October 31, 1998.

[16] Michael Willis, "The Islamist Movements of North Africa," in *Security Challenges in the Mediterranean Region*, Roberto Aliboni, George Joffé, and Tim Niblock, eds. (London, and Portland, Oreg.: Frank Cass, 1996), 22.

[17] Andres Ortega, "Relations with the Maghreb," in *Maelstrom: The United States, Southern Europe and the Challenges of the Mediterranean*, John W. Holmes, ed. (Cambridge, Mass.: The World Peace Foundation, 1995), 40.

[18] See, for example, NAA, Civilian Affairs Committee, Sub-Committee on the Mediterranean Basin, *The Rise of Islamic Radicalism and the Future of Democracy in North Africa*, AL 199 CC/MB (94) 4 by Mr. Pedro Moya (Spain), Rapporteur (Brussels: International Secretariat, November 1994).

[19] *Assembly of Western European Union, Proceedings, 42nd Session, December 1996, I. Assembly Documents* (Paris: WEU), Doc. 1543, November 4, 1996, "Security in the Mediterranean Region." Report submitted on behalf of the Political Committee by Mr. de Lipkowski, Rapporteur, 169; and, Alvaro Vasconcelos, "General Framework and Concepts," in *Security in Northern Africa: Ambiguity and Reality*, Fernanda Faria and Alvaro Vasconcelos (Paris: Chaillot Papers 25, Institute for Security Studies, WEU, 1996), 13.

[20] For a background of developments in Algeria in the 1990s, see William H.

Lewis, "Algeria at 35: The Politics of Violence," *The Washington Quarterly* 19, 3 (summer 1996): 3–18; and, Hugh Roberts, "Algeria's Ruinous Impasse and the Honourable Way Out," *International Affairs* 71, 2 (April 1995): 247–67.

[21]Nicole Grimaud, "Tunisia: Between Control and Liberalization," *Mediterranean Politics* 1, 1 (summer 1996): 97.

[22]NAA, Civilian Affairs Committee, Tunisia Report, 1995, 6–8.

[23]*Keesings Record of World Events 1994*, 40161.

[24]Murat Terterov, "Egypt: The Islamist Challenge," *Mediterranean Politics* 1, 2 (autumn 1996): 246.

[25]Ibid., 247.

[26]Willis, "The Islamist Movements of North Africa," 17–19.

[27]Fernanda Faria, "Security Policies and Defence Priorities," in *Security in Northern Africa*, 43.

[28]Pazzanita, "Political Transition in Mauritania," 44–45.

[29]For more details of the IAF in Jordanian politics, see Robinson, "Can Islamists Be Democrats?" 373–87.

[30]An observation made by Willis, "The Islamist Movements of North Africa," 23–24.

[31]Saad Djebbar, "The Algerian Presidential Election and Its Consequences," *Mediterranean Politics* 1, 1 (summer 1996): 123.

[32]Lewis, "Algeria at 35," 10.

[33]Faria, "Security Policies and Defence Priorities," 44, n.86; and Pazzanita, "Political Transition in Mauritania," 49.

[34]Rémy Leveau, "Morocco at the Crossroads," *Mediterranean Politics* 2, 2 (autumn 1997): 104.

[35]Cassandra, "The Impending Crisis in Egypt," 21–24.

[36]Faria, "Security Policies and Defence Priorities," 23.

[37]Roger Owen, *State, Power and Politics in the Making of the Modern Middle East* (London and New York: Routledge, 1992), 208–09.

[38]Saad Eddin Ibrahim, "Crises, Elites, and Democratization in the Arab World," *Middle East Journal* 47, 2 (spring 1993): 303.

[39]Alan Richards, "Economic Imperatives and Political Systems," *Middle East Journal* 47, 2 (spring 1993): 217–27.

[40]Augustus Richard Norton, "The Future of Civil Society in the Middle East," *Middle East Journal* 47, 2 (spring 1993): 205–16.

[41]Alejandro Colas, "The Limits of Mediterranean Partnership: Civil Society and the Barcelona Conference of 1995" (paper presented at the International Studies Association Convention, Toronto, March 19–22, 1997), 12–13.

[42]Tim Niblock, "A Framework for Renewal in the Middle East?" in *The Middle East in the New World Order*, 10–12.

⁴³Rémy Leveau, "The Future of the Maghreb," in *Peace and Stability in the Middle East and North Africa,* Josef Janning and Dirk Romberg, eds. (Gütersloh: Bertelsmann Foundation, 1996), 105.

⁴⁴Ibrahim, "Crises, Elites and Democratization in the Arab World," 297.

⁴⁵NAA, Civilian Affairs Committee, Tunisia Report, 1995, 4–5.

⁴⁶Joffé, "Elections and Reform in Morocco," 225.

⁴⁷Vasconcelos, "General Framework and Concepts," 10.

⁴⁸Michael Köhler, "Stability in Algeria, Morocco and Tunisia," in *Peace and Stability in the Middle East,* 116; and Leveau, "Morocco at the Crossroads," 103.

⁴⁹Joffé, "Democracy in the Maghreb," n.5, 36, 39, 58.

CHAPTER 5

The Southern Mediterranean and the Middle East in the Post–Cold War Era: Economic and Social Issues and Perceptions of Security

INTRODUCTION

As in the previous chapter, the focus here will be on the non-NATO Mediterranean states that are participating in the Mediterranean Initiative. Algeria will also be included in the analysis because of the significance of recent developments there for other neighboring and nearby states. Again, the case of Israel will not be discussed specifically. Economic and social problems and security issues are closely interlinked concerning Arab states in the Mediterranean area. This chapter will focus on how ruling elites in these states perceive security, and how their perceptions may differ from those held by members of their publics. In this particular context, it would seem that Israel is bracketed together with states north of the Mediterranean. Certainly, Arab governments and peoples are fully aware of Israel's traditionally close military ties with the West and, in particular, Israel's strategic alliance with the United States.

The economic problems that governments in North Africa and the Middle East have to confront are serious and manifold. In many cases, the transition from what was largely a state-controlled-and-planned economy to one based more on the free market is proving to be a painful experience. Privatization needs to be encouraged, outside investment must be attracted, substantial debts have to be paid off and budget deficits slashed. Economies ought to become more diversified, especially those that customarily have been dependent on oil exports and thus have been subject to the fluctuations in prices of oil. In some cases, less dependence on agriculture and more industrialization programs could be promoted. The debt problem will not ease unless military expenditures are reduced. Demographic problems also must be considered. High birth rates will exacerbate the current situation of overpopulation and soaring levels of unemployment, especially among the youth. Discontented youth are more likely to turn to radical, politicized Islam in the hope that their grievances will be addressed. Overpopulation may contribute to environmental problems as land is exhausted in a desperate effort to feed the population and overcrowded cities become increasingly more polluted. In these circumstances, improved access to the EU market and more assistance from the EU could help offset, at least partially, some of these difficulties.

The debt problem is a result of a high degree of economic mismanagement, corruption, and unsuccessful attempts to modernize the economy. According to World Bank data, in 1994 the total foreign debt of certain North African and Middle Eastern countries was as follows: Egypt, $33,358 billion; Algeria, $29,898 billion; Morocco, $22,517 billion; Tunisia, $9,254 billion; Jordan, $7,051 billion; and Mauritania, $2,326 billion.[1] Servicing these debts around that time totaled 71 percent of the exports of goods and services in the case of Algeria, 30 percent in Morocco (down from a recent figure of 60 percent), and 25 percent in Tunisia.[2] These states are hence encountering serious problems in their efforts to encourage foreign investment. In addition, political risks, relatively high labor costs, and widespread corruption mean that a return on investment will be low. In recent years there have been attempts to modify laws and regulations to attract outside capital. For example, in some instances foreigners may purchase up to 50–51 percent of shares in companies. However, with the exceptions of Morocco and Tunisia, foreign investment in North Africa and in the Middle East has actually been dwindling.[3]

THE PROBLEM OF MIGRATION

There has been much talk in the West, and in NATO circles, about how economic and social tensions could contribute to the spread of radical, politicized Islam in North Africa. This spread could trigger a wave of mass migration from the southern to the northern shores of the Mediterranean that could threaten the internal security of certain NATO member states. This issue thus warrants some attention. There are already a substantial number of registered residents from Algeria, Morocco, and

Tunisia living in parts of Europe. In 1994, approximately 1,393,000 registered residents from these states lived in France. Another 164,500 were registered in the Netherlands, 160,000 in Belgium, 134,000 in Italy, 130,000 in Germany, and 67,500 in Spain.[4]

More people from the southern Mediterranean are likely to enter Europe by legal or illegal channels so as to find employment unless there is an improvement in the economic situation in North Africa. Extreme right-wing groups in France, Italy, and Germany, in particular, have voiced their concerns that further migration could disturb the balance in their societies. They have referred to the danger of Islamic fifth columnists aiming to destabilize Europe. This threat to the societal security of the West appears to be somewhat exaggerated. Islamic communities based in Europe did not create problems at the time of the Gulf War after the Iraqi invasion of Kuwait, or throughout the Bosnian crisis.[5] On the other hand, Algerians based in France were behind a spate of terrorist acts in Paris, most notably in 1994 and 1995.

In the 1960s, governments in western Europe were eager to employ migrant workers from abroad.[6] In particular, the economies of Belgium, France, the Netherlands, and West Germany were suffering from a shortage of labor in various, mainly unskilled jobs. At the same time, governments in North Africa and in the Middle East were keen to allow their workers to be employed overseas in order to bring back remittances and acquire some basic skills. Arab governments, therefore, encouraged migration in order to ease unemployment problems at home and improve their balance of payments. Initially, governments north and south of the Mediterranean believed that these migrant workers would eventually return to their native countries. In general, this did not occur. The oil crisis of 1973 and the economic recession at that time, and the growing concerns in Europe at the social impact of immigration in a period of rising unemployment, resulted in a freeze on labor recruitment in western Europe. Often in vain, Western governments offered incentives to encourage the migrant workers to return to their home countries. A new wave of migration from North Africa targeted Italy and Spain, where there were few immigration controls and where economic growth demanded access to cheap labor. Tunisians and Moroccans sought employment in Italy and mainly Moroccans entered the labor market in Spain.

By the 1980s there was also increasing pressure from so-called asylum seekers from North Africa who claimed that they were the victims of persecution at home and were hoping to settle in Europe. In 1985, for the first time, the number of asylum seekers—whether genuine or not—

exceeded the total of legally admitted foreign workers in Europe. According to Gil Loescher, a new era had commenced.[7] As avenues for legal immigration were tightened, there was an increase in the number of illegal immigrants. Some immigrants were simply seeking a better life and were not the victims of political persecution. By the early 1990s there were further efforts to tighten up the EU's borders as governments in western Europe, confronted again with recession, also feared a possible mass wave of migration from post–Cold War eastern Europe (which never actually materialized). EU members were also aiming to harmonize their migration policies with a view to suppressing their internal borders and creating a single market. The outbreak of serious civil unrest in Algeria in 1992 resulted in renewed efforts by western European governments to curb migration. In September 1990 and May 1991, the Italian and Spanish governments, respectively, introduced obligatory visa requirements for Maghrebi nationals.[8] In the aftermath of Operation Desert Storm, Maghrebi nationals were also no longer able to send workers to oil-rich Gulf states to secure remittances. Countries such as Kuwait preferred to recruit labor from Asia.[9] It was believed that Asian workers were much less likely to become involved in the internal politics of the Gulf states.

Demographic problems in North Africa will not disappear in the foreseeable future. A continued high population growth will result in many people still seeking to migrate northward to secure a better future. At present, the average annual growth rate of the population in North Africa is 3 percent. In Algeria the population swells by almost one million each year. The figure for Egypt is even higher at 1.3 million. The population in North Africa and the Middle East is projected to double from 240 million in the early 1990s to almost 500 million by the year 2020.[10] The population immediately north of the Mediterranean will only increase at an annual rate of less than 0.3 percent over the same period. This will lead to the size of the population in the southern and eastern Mediterranean outstripping that of the northern Mediterranean within this time period.[11] Workers, and often families already settled in Europe, act as "important bridgeheads." They are a source of attraction for the migration of other workers and dependents from their original village or region.[12] It is in those countries where Islamic communities have been previously established that there is often the greatest determination to resist further immigration. These communities at times create problems when resisting attempts by the local authorities to assimilate them.[13]

It is not clear whether there will be a continued need in western Europe in future for unskilled and semiskilled labor. The traditional markets for unskilled foreign labor are shrinking with the advance in automation.[14] French demographers, though, have predicted that there will be a 30 percent shortfall in indigenous European labor supply by 2025.[15] Youth from the Maghreb in search of work could compensate for Europe's increasingly aging population.[16]

Joffé has argued that the Europeans are acting illogically. According to him, given that the demand for cheap migrant labor is expected to top 56 million by the year 2100, it makes far more sense to meet this demand through legal, regular channels than to become dependent on illegal migrants as is increasingly the case.[17]

There is certainly a need for governments north and south of the Mediterranean to cooperate more with regard to the migration problem. Obviously, the Europeans are in favor of imposing some form of control and regulation concerning the number of North Africans seeking entry to Europe. It is also not in the interests of Arab governments to lose large numbers of its young and potentially dynamic workforce even if they send remittances back home for some time. Many of these workers and their families may never return to North Africa.

Habib Ben Yahia, when foreign minister of Tunisia, spoke of how imperative it was for the North Africans and the EU to manage jointly the migration problem. Ben Yahia himself offered no specific recommendations.[18] One possibility is to prepare a General Agreement on Migration that would make use of quotas. This agreement could complement a General Agreement on Human Resources Policy. The latter agreement could stress the need for improved education and training.[19] It will not be easy, though, to allocate quotas according to the needs of the market and society.[20] An all-encompassing general agreement is unlikely. More probably, bilateral deals will be concluded similar to those already negotiated. For example, in February 1992 Spain and Morocco signed a "readmission agreement." The Moroccan government committed itself to readmit all undocumented immigrants apprehended in Spain who originated from or had traveled through Morocco. By the end of 1992 the Moroccan government had stepped up surveillance of its northern coast. Similar readmission agreements were apparently concluded by the French and Algerian authorities in 1993 and 1994.[21]

Another proposal for handling the migration problem is to simply throw money at it in the hope that it will disappear. According to this argument, if the EU were to increase its economic aid and assistance programs to North Africa and the Middle East, improved living and working conditions at home would then discourage many potential immigrants from seeking to start a new life in Europe. More foreign investment in the Arab economies would be required. In order to resolve the unemployment problem labor-intensive industries in North Africa should be encouraged. European markets should be opened to labor-intensive manufactured goods and agricultural produce. Simultaneously, Arab markets should be opened to European imports in order to overcome the shortage of various items. This shortage has also driven many to migrate northward across the Mediterranean. Income differentials between states north and south of the Mediterranean should be considerably reduced.[22] Should aid, however, be given to a repressive government? For example, the US administration has not welcomed extensive French financial support for the Algerian authorities.[23] Again, one returns to the problem of so-called conditionality, noted in the last chapter. If Western governments impose conditions before aid and financial assistance are dispatched to a particular state, the West is then leaving itself open to the charge of seeking to interfere in the internal affairs of Arab states.

The migration issue also has other dimensions. There are environmental refugees, the products of ecological degradation. With overpopulation, many are unable to work on the land. Marginal lands are then overexploited leading to erosion. Trees are felled to make way for new land, but this often only accelerates the process of desertification. Urban centers become overcrowded and polluted. Clean water becomes a scarcity.[24] Migration movements from sub-Saharan African must also be considered. In recent years France has hosted 150,000 sub-Saharan immigrants. Italy has taken a further 50,000, and Portugal around another 30,000. There is also a south-south aspect to this problem. Some people from sub-Saharan Africa intending to go to Europe decide to remain in North Africa when supposedly transiting through the Maghreb. The further tightening of Europe's borders has contributed to this trend.[25]

Finally, any possible large-scale migration from Algeria, if the civil unrest there continues to escalate, may lead to a flood of people entering Morocco and Tunisia, with potentially serious destabilizing consequences for the authorities in Rabat and Tunis.[26]

On the other hand, it may be argued that the migration problem has been overexaggerated. Perhaps it is unlikely that the establishment of some form of Islamist government in a particular state in North Africa will trigger a mass migration from that state to Europe or to another neighboring Arab state. It would seem that only the Westernized and wealthy ruling and financial elites would seek to flee the country. Some businessmen would remain in the hope of playing an influential commercial role for the new government. The majority, being the poorer sections of society, would presumably also stay, in the belief that Islam could address their grievances.[27]

It appears that the governments in North Africa and the Middle East have to an extent played the migration card in order to receive more attention and support from Germany and other states in northern Europe. The northern Europeans have been less interested in developments in the southern Mediterranean.[28] In recent years, though, the EU as a whole has taken an enhanced interest in the Maghreb, and in the Mediterranean more generally. And NATO has also assumed more of an interest in Mediterranean affairs after the end of the Cold War.

Western governments have to tread carefully with regard to the migration issue. Tough, discriminatory visa regimes aimed at the Maghreb would be perceived in the south as shameful and racist. Opposition parties and trade unions in North Africa are already accusing Western governments of adopting such unfair policies.[29] If confidence-building measures between governments and peoples north and south of the Mediterranean are to develop, the migration problem must be tackled in a way satisfactory to all. Obviously, there must be some form of regulation to channel migration flows, but these regulations should be jointly worked out by the relevant authorities north and south of the Mediterranean. A more economically stable if not prosperous south is also in the interests of both the Europeans and the peoples of North Africa and the Middle East.

The migration issue is being tentatively addressed in such forums as the EU's Barcelona Process and the Mediterranean Forum. NATO's Mediterranean Initiative has steered clear of directly addressing this issue, which is clearly not an area of competence and expertise for the Atlantic Alliance. However, NATO policy-makers are fully aware of the possible negative consequences for the security and stability of the Mediterranean if the migration issue is not successfully tackled. Officials at NATO are thus presumably following with interest and attention how the Barcelona Process, for example, is seeking to resolve the long-standing migration problem.

THE CURRENT ECONOMIC SITUATION

It was serious economic problems by the mid-1980s—with corruption, mismanagement, and the fall in oil prices—that led to increasing civil unrest in Algeria, culminating in the military imposing a state of emergency in 1992. Algeria was an example par excellence of a country overdependent on the export of one or two products, in this instance oil and gas. There is a need to diversify the economy. In spite of the lack of genuine democratization in the country, fear of the possible seizure of power by radical Islamist elements in Algiers has resulted in Algeria benefiting from Western financial backing in recent years. In particular, the French government, eager to avoid a further spillover of the Algerian problem onto French territory, has extended bilateral aid to the tune of approximately $1.2 billion per annum.[30] French officials also succeeded in securing a favorable rescheduling of Algeria's debt and convinced the IMF to offer support. Thus, in May 1995 the IMF approved a $1.8 billion loan to assist the Algerian government in its restructuring and privatization program. However, oil and gas reserves can still be important revenue earners. In addition to the Trans-Mediterranean gas pipeline that extends to Italy, in November 1996 the new Maghreb-Europe pipeline began to pump gas from Algeria to Spain and Portugal via Morocco. Companies such as BP are prepared to invest huge sums of money to develop the considerable natural gas reserves in Algeria. By the mid-1990s there were signs of an improvement in the Algerian economy as GDP began to rise and inflation fell below 20 percent. However, for Algerians to benefit from a real economic takeoff, it would seem that political stability and peace should first be restored.

In Egypt, continued economic hardship must be a source of concern for the leadership. Since 1991, under IMF supervision, inflation has been reduced to single-digit figures, the budget deficit has been lowered, and foreign reserves have increased as tariffs were cut. Further privatization is planned. A government-sponsored social-development fund has provided support for the unemployed. However, it seems that in contrast to the macroeconomic performance, at the microeconomic level problems must still be addressed. High prices, low wages, and the shortage of jobs do not augur well. Much more privatization is required together with further trade liberalization in order to encourage foreign investment.[31] By the mid-1990s, according to official figures, about 20 percent of the labor force lay idle. With a rapidly growing population, the situation is likely to deteriorate, and it is projected that within a few years up to six million Egyptians will probably be living on less than one US dollar a day.[32] Nevertheless, by autumn 1997 there were reports that the privatization and liberalization program was starting to pick up. Land reform meant that controls on the fixing of rents for farmland were reduced, although in the short term this would seriously disadvantage the one million or more tenant farmers. And the amount of foreign investment in the Egyptian economy was on the increase.[33]

The Jordanian economy also is in trouble. Riots erupted in August 1996 following food price hikes in accordance with an IMF-imposed austerity program that aimed to reduce Jordan's substantial debt. A curfew was enforced in several cities and King Hussein suspended parliament for a period. These had been the worst riots since 1989 when street disturbances had followed a rise in fuel prices. The Jordanian economy has suffered as a consequence of Saddam Hussein's attempted conquest of Kuwait and the imposition of UN economic sanctions against the Baghdad government. Iraq had been Jordan's major trading partner. No longer able to barter for Iraqi oil, the Jordanians are forced to purchase more expensive oil from Syria. There has also been the loss of remittances after the expulsion of Palestinians from Kuwait following its liberation and the forced removal of Palestinians from other Gulf states. Many of these Palestinians possessed Jordanian passports and had channeled their funds through Amman. The United States has had to increase its financial support to Jordan. There is the possibility that the Jordanian economy will benefit from increased trade turnover with Israel with the opening of borders after the signing of the Israeli-Jordanian peace treaty in 1994.[34]

The imposition of VAT on foodstuffs led to a 25 percent increase in the price of bread that sparked riots in the Mauritanian capital of Nouakchott in January 1995. A curfew was announced and a number of opposition leaders arrested at the time. In contrast to other Arab states in North Africa and the Middle East that have reserves of oil and gas or are endowed with other raw materials, Mauritania is a poor country. It has few natural resources to exploit, apart from deposits of iron ore and a largely undeveloped fisheries sector. However, in the 1990s there has been a sustained economic growth after the government adopted an IMF structural-adjustment program.[35]

Serious economic reforms started to be implemented in Morocco after September 1983 when the government was unable to meet the schedule with regard to the repayment of a foreign debt of around $13 billion and was compelled to turn to the IMF for support. Previously, the Moroccan economy had thrived due to large increases in the price of phosphates but by the early 1980s the boom had ended. Following IMF-prescribed measures, subsidies in consumer goods were cut and a liberalization of the trade regime commenced. As the gap between the rich and poor widened there was civil unrest in 1984 and 1990. Nevertheless, the economy grew on average by 5 percent each year in the period 1985–91. More problems arose after a series of droughts in 1992–93 and 1994–95. The Moroccan economy was too dependent on the success or failure of its agricultural production. By the mid-1990s unemployment reached 20 percent and the population was continuing to expand rapidly. Considerable imports of food were needed to satisfy the population's needs. There were attempts to develop textile and leather industries but more outside investment was required to improve basic infrastructure and modernize the education system. Further privatization of the economy was needed.[36]

In spite of its largely repressive regime, the Tunisian economy has fared reasonably well. Although oil reserves are declining new sources of gas have been discovered. The economy has grown on average 4 percent each year since 1970. More significantly, in contrast to its neighbors, there has been a decline in the birth rate. By the 1990s the population of Tunisia totalled less than nine million. As a consequence, income distribution is relatively even. Boasting an income of $1,800 per capita, the Tunisian economy was the most successful of all the Maghreb states. A sizeable middle class was beginning to form. The economy has also benefited from improvements in the manufacturing industry and a better network of irrigation for agriculture.[37] Perhaps the continued development of the Tunisian economy may lead to less political repression at home, as the threat of radicalized Islam recedes. Certainly, Tunisia is performing far better than its neighbors in the economic sphere. The IMF and World Bank often cite Tunisia as an example of a country that has successfully adjusted its economy. However, even the Tunisian economy is still likely to suffer if a Mediterranean free-trade area—examined later in this study—is introduced in 2010.[38]

TRADE ISSUES

There is an interdependent relationship between the EU and North Africa and the Middle East concerning trade. The EU's interest in the oil and gas reserves of the Maghreb has been noted. There is also, to some extent, a European dependence on cheap labor from North Africa, which could be reinforced in the next century. According to IMF figures, in 1995, exports from the EU to the Middle East and North Africa totalled $77.5 billion or 18 percent of all EU exports to the developing world. Imports from North Africa and the Middle East to the EU amounted to $58 billion or 15 percent of the EU's total imports from developing countries.[39]

The EU, though, is not crucially economically dependent on the Arab world. Rather, Arab states are critically dependent on western Europe. For each North African state the proportion of their exports to the EU has risen substantially since 1985. By 1992 western Europe received 57.5 percent of Egyptian total exports, 67 percent of Moroccan exports, 74.8 percent of Algerian exports, 75.6 percent of Tunisian exports, and 87.2 percent of Libyan exports.[40] Europe has also been an important supplier of aid. The total aid disbursed by the EU—both collectively and bilaterally—to countries around the southern and eastern Mediterranean totaled $2.8 billion in 1994.[41] Arab governments have clamored for more assistance from the EU. They complain that the countries of central and eastern Europe have been receiving more aid from the EU since the end of the Cold War.

Over the last decade or so the North African states have been disadvantaged with regard to their commercial dealings with the EU/European Community. The admission of Portugal and Spain to the EC in 1986 meant that the EC had become self-sufficient in agricultural produce. Consequently, quota restrictions were imposed on fruits, vegetables, olive oil, and wine from Algeria, Morocco, and Tunisia. After the single European market came into operation on January 1, 1993, a series of nontariff barriers—quotas and other protectionist measures—were raised against non-EC imports. Imports to the Community had to meet new European industrial standards. Much investment would be required for the North African economies to improve the quality of their goods to this level. Morocco and Tunisia, in particular, raised objections.[42]

The EU is likely to remain a major trade partner for the states of North Africa and the Middle East for the foreseeable future. There is little intraregional trade between the Arab economies of North Africa. The Arab Maghreb Union (AMU), established in the late 1980s, whose members are Algeria, Libya, Mauritania, Morocco, and Tunisia, has failed to achieve economic and political integration in North Africa. Originally, AMU members had planned to create a common market by the year 2000. They had also aimed to form common institutions such as a joint consultative parliament and an investment bank.[43] The economies of AMU member states, however, are not complementary. Indeed, they are actually trade competitors. By the mid-1990s intra-AMU trade amounted to only 3 percent of AMU member countries' total foreign-trade turnover. For the same reasons, moves initiated in 1998 to establish an

Arab Free Trade Area (AFTA) by 2008 are also not likely to boost intraregional trade in North Africa. The envisioned free-trade regime is replete with exceptions, including trade in agricultural goods.[44]

Relations between the EU and the Arab states of North Africa and the Middle East are likely to further expand as part of the EU's Barcelona Process. The goal is to achieve a free-trade area in industrial goods and services throughout all of the Mediterranean by the year 2010. The significance of the EU's Barcelona Process will be discussed later in this study. As a military and political organization, it would seem that NATO can do little to tackle the economic problems confronting governments in the southern and eastern Mediterranean. These economic difficulties could heighten political instability in the Mediterranean if not successfully addressed in the near future. However, CBMs, encouraged and even developed by NATO in the Mediterranean as a key element of its Mediterranean Initiative, could help to ease tensions in the area. These CBMs may allow Arab governments to devote more of their resources to economic and social problems instead of building up larger arsenals of conventional weapons and WMD.

SOUTH-SOUTH SECURITY ISSUES

It was noted in the first chapter that as well as having security concerns with regard to relations with NATO member countries and other outside states, Arab governments in North Africa and the Middle East also focus their attention on relations with the governments of neighboring states. Inter-Arab relations are far from smooth. In this context, classic, hard security issues are often on the agenda of Arab governments. Although there has been much discussion concerning the proliferation of WMD in the southern Mediterranean and their possible use against NATO member states, these same weapons would more likely be deployed against fellow-Arab countries and against Israel.

The arms race in the Middle East is motivated primarily by "intraregional concerns." There is no systematic enmity to the north, in general.[45] A war between states in the southern Mediterranean would usually require ballistic missiles of only relatively short range compared to the types of more advanced weaponry needed to strike western Europe. Concerning WMD specifically, in addition to their deterrent value the possession of such weapons is a mark of prestige for a developing state. There is much value placed on the actual nonuse of these weapons.

Libya is a pacesetter in North Africa and the Middle East concerning the acquisition of WMD. Other states in the area have been compelled to attempt to match the expanding Libyan arsenal. By the mid-1990s Libya had already obtained SCUD-B medium-range missiles with a range of 280–300km, and upgraded SCUD-C missiles from North Korea with a longer range of 500km. Libya was also apparently interested in procuring Chinese M-9 missiles with a range of 600km, and Chinese CSS-2 missiles able to strike a target 2,800km away.[46] The Libyan government has expressed an interest in securing the North Korean No-Dong 1 missile, which has a range of 1,000-1,300km. This missile could carry a nuclear, chemical, or biological warhead.[47] Apparently there are large chemical-weapon production sites in Libya. Chad has already accused Libya of using chemical weapons in a war fought between the two states in 1986–87. Perhaps Libya had made use of missiles acquired from Iran in that conflict, because at that time Libya had not developed its own chemical-missile warhead capability.[48] Neighboring Arab states must be wary of the size and capability of Libya's military machine under the control of the unpredictable Gadhafi. And, of course, NATO member states are also closely monitoring Libya's military assets, mindful of Gadhafi's strike against the Italian island of Lampedusa in 1986. However, it appears that Libya will not have a biological or nuclear weapon capability for the foreseeable future.

Syria and Iran have also, it seems, acquired SCUD-C missiles that could be targeted against Israel and Turkey, a NATO member state. It appears that using Russian technical assistance, Iran is developing the Shihab-3 missile (range 1,300km) and the Shihab-4 missile (range 2,000km). Unlike Syria, Iran may well have pretensions toward developing a nuclear weapons program. Iraq may have been able to hide a few SCUD-C missiles from UN arms inspectors. Precise details of possible chemical and biological weapons programs in Iran, Iraq, and Syria are difficult to ascertain. Algeria, apparently, has a ballistic missile capability but has no ambitions to develop biological or chemical weapons. The Chinese-built nuclear reactor in Algeria is now under international safeguards. Egypt had employed chemical weapons as early as the 1960s when involved in the Yemeni civil war. Most probably Egypt has developed its chemical-weapons capabilities over the following decades.[49] Israeli officials are aware of Egypt's potential to wage chemical warfare.

Certainly, the disagreements between Israel and Egypt concerning WMD need to be resolved as quickly as possible in order to boost the prospects for enhanced stability in the Mediterranean. The positive developments in Arab-Israeli relations in the early 1990s had originally raised hopes that the arms build-up in the area would be slowed. Differences of opinion between Israel and Egypt and other Arab states almost held up the extension of the Nuclear Non-Proliferation Treaty (NPT). These disagreements have been the main reason for the freezing of talks in the multilateral working group known as ACRS, which had been set up to help further the Middle East peace process. Arab states were opposed to limitations being placed on their conventional weapons and were against attempts to hinder their chemical-weapons programs when Israel had already acquired a nuclear weapons capability and had refused to sign the NPT. Egypt has evidently redoubled its efforts to develop ballistic missiles that may be tipped with chemical warheads.[50] Egyptian SCUD-B missiles have been modified to extend their range from 280 to 450km. The Badr missile is also being developed, which could have a range of 850-1,000km.[51] Israel has deployed Jericho-1 and Jericho-2 missiles with a range of 650km and 1,500km, respectively. Clearly, this escalatory arms race in the Middle East is inherently destabilizing.

It is important to note that NATO member states themselves are responsible for the sale of 80 percent of all weaponry to the Middle East. The United States, Britain, and France, in particular, are major arms suppliers.[52] NATO could address this issue of arms transfers in the near future, perhaps within the context of its Policy Framework on the Proliferation of Weapons of Mass Destruction, which is discussed in the next chapter.

Given the proliferation of various weapons in the area, it is fortunate that a number of border disputes between states in the southern Mediterranean have been resolved, but potential sources of conflict remain. The problem over the future of the Western Sahara had originally led to friction in relations between Algeria, Mauritania, and Morocco. By 1979 Mauritania, had relinquished its claim to the disputed territory. Algeria had continued to support POLISARIO guerrillas, who resisted Moroccan occupation forces in the Western Sahara. The government in Algiers also diplomatically recognized the so-called Sahrawi Arab Democratic Republic as the ruling authority over the Western Sahara even though Morocco had established a control over the territory by the time the republic was declared in 1976. Moroccan officials suspect that Algeria is seeking an outlet to the Atlantic Ocean via the Western Sahara. In recent years Algerian officials have been distracted by internal problems and have scaled down their support for POLISARIO. The UN has been struggling to organize a referendum among the Sahrawi people in the Western Sahara in an endeavor to solve the issue according to the principle of self-determination. If the UN fails to obtain a peaceful solution, the Western Sahara could become the scene of a major confrontation between Algeria and Morocco if and when the civil unrest in Algeria is eventually brought under control.

Morocco and Mauritania have yet to settle their differences over the status of the port of La Gouera and the security of the rail connection between Zouirat and the Atlantic Ocean. Zouirat was traditionally an important center for iron mining in Mauritania, although by the 1990s iron exports had declined and made up only 17 percent of Mauritania's GDP.[53] However, a military engagement between Mauritania and Morocco is unlikely given the far greater size of the Moroccan armed forces. Mauritania will more probably be involved in border skirmishes with its southern neighbors, Mali and Senegal. In the mid-1990s approximately forty thousand Tuaregs and Moors from Mali had sought refuge across the border in Mauritania. At the same time, fifteen thousand black Africans from Mauritania had fled over the border to Mali and another fifty thousand were still in Senegal after having been expelled by the Mauritanian military government in 1989. These local disputes are connected with the politics of Saharan and sub-Saharan Africa and are not about to impact on relations between states to the immediate north or south of the Mediterranean.

Egypt and Sudan were unable to reach an agreement over the strategically important territory known as the Halaib triangle by the Red Sea. Egypt finally seized the area of land in 1995. This came after Egypt accused the Sudanese of attempting to assassinate President Mubarak at the summit of the Organization of African Unity (OAU) in Addis Ababa. Egyptian officials have frequently charged that Sudan is providing training on its territory for radical Egyptian Islamic groups that are seeking to overthrow the Mubarak government by violent means.[54]

The failure of states in the southern Mediterranean to cooperate with one another with regard to their security concerns is well illustrated in the lack of progress in developing the AMU. The AMU aims to coordinate the activities of member states in the economic, cultural, and defense spheres. Difficulties in the economic field have already been noted. Cooperation in defense matters is virtually nonexistent. By the terms of Article 27 of the Treaty of Marrakesh in 1989—the treaty that established the AMU—member states had pledged to bind themselves together in mutual defense. This was, in effect, a nonaggression pact. In January 1990 a Defense Council within the AMU was formed. In practice, the AMU became little more than a talking shop. Member states squabbled over the goals and objectives of the AMU. Libyan officials advocated a full, political union. Morocco and Tunisia preferred closer economic integration. Algeria argued for a looser form of association. There was a lack of consensus over whether Egypt should be invited to join, thereby giving the AMU a more Middle Eastern focus. The Western Sahara problem also created tensions within the ranks of the AMU. Additionally, member states were at loggerheads over how to react to the Iraqi invasion of Kuwait. The domestic unrest in Algeria, and the imposition of UN sanctions on Libya after the Lockerbie disaster also created divisions among AMU members.[55] The AMU is unlikely to be a stabilizing force in the southern Mediterranean in the foreseeable future with this track record.

As noted in the first chapter, south-south disputes may have an impact on north-south relations in the Mediterranean area. In the immediate aftermath of the Gafsa incident in 1980 along the Tunisian-Libyan border, the French fleet had been poised to come to the defense of Tunisia in the event of a full-scale Libyan assault. Various NATO member states—given their interest in continued access to North African oil and gas and their concern for the safety of Western civilians living and working (often on oil and gas projects) in the southern Mediterranean—are maintaining a close eye on potential south-south disputes. A breakdown in the Middle East peace process could have wider repercussions that would affect NATO's relations with the southern and eastern Mediterranean. It is clearly not realistic to expect that as a direct result of their participation in NATO's Mediterranean Initiative the six non-NATO Mediterranean states will be able to settle between them their differences. Nor is it at all likely that these six states will call upon NATO to act as some type of intermediary to help mediate and settle their disputes. However, if, as a result of its Mediterranean dialogue, NATO is able to build confidence between the governments and peoples of states north and south of the Mediterranean, this could have a beneficial impact on south-south relations in the area. Clearly, though, the Arab-Israeli problem will only be resolved if much more progress is made in the separate and ongoing Middle East peace process.

SOUTHERN PERCEPTIONS OF SECURITY

A previous chapter covered the various definitions of security, noting how since the end of the Cold War there has been a tendency to refer more to an enlarged concept of security in which the military aspect is only one of several other possible dimensions. The notion of a shared common or cooperative security between states is problematic in the Mediterranean area because of the different conceptions of security held by groups north and south of the Mediterranean. Cultural differences must be taken into account. But within the Arab world, governments may not perceive their security interests similarly, and also Arab ruling elites may not share the same security concerns as their general publics. The focus of this section is on the various perceptions of security held by certain officials and commentators in states south of the Mediterranean and how the "south" in this context perceives the "north." Northern perceptions of a possible security threat from the south will be examined later in this study.

Clearly, governing elites in North Africa and the Middle East busy themselves with traditional security concerns as seen in their efforts to procure conventional weapons and possibly also WMD that could be used against Israel or their Arab neighbors. These same elites are also apprehensive about the threat to their security from Islamist or other opposition groups within their state. Governing officials should remain sensitive to how their publics may react to agreements concluded with outside states and in particular with Western states and organizations. In general, in contrast to their publics, Arab leaders are less suspicious of the policies of the West. Nevertheless, even Arab ruling elites remain convinced that the West at times seeks to interfere in their internal affairs. These elites may perceive the proclaimed interest in furthering democratization as a Western attempt to impose its own system of values on another culture and society. The Cold War, with its East-West confrontation, was regarded by ruling Arab officials as a kind of "assurance" against Western domination. In the post–Cold War there are less constraints on Western decision-makers. Arab governments find it difficult to trust organizations such as NATO because they perceive the West to be anti-Arab and anti-Muslim. They were critical of the reluctance of Western governments to become involved in Bosnia in contrast to the immediate response to the Iraqi invasion of Kuwait, and they were unhappy at the determination of Western officials to maintain UN economic sanctions against Libya.[56]

The views of Habib Ben Yahia, the then foreign minister of Tunisia, in an article published in 1993, are worth noting. These were the opinions of a leading politician of a country that is traditionally close to the West, and to France in particular. According to Ben Yahia, states south of the Mediterranean do not take kindly to how the Europeans constantly refer to the south as the source of threats and problems stemming from weapons proliferation, migration, extremist Islam, cultural differences, overpopulation, and so forth. In Ben Yahia's opinion, the north regards the southern Mediterranean as an asset over which it must dominate. In order to legitimize this dominance the north depicts the south as a menace. In turn, as a consequence, the south is provoked and looks upon the north as a disdainful and rich club that is sealing its borders to build a European social and economic fortress.[57] There is an assumption here that the north regards itself as superior—culturally superior? Ben Yahia does not make clear how the north will be able to dominate the south if it is seeking to close its borders. It would seem that cultural domination would require a much more visible northern presence in the south. The argument of Ben Yahia echoes that of other Arab officials who complain that the West is not paying enough attention to the security problems of the south.

Constant references in the north to a possible military threat from south of the Mediterranean have the danger of becoming self-fulfilling prophecy. They foster certain perceptions of a north-south divide.[58] The Tunisian academic Khalifa Chater, referring to the Mediterranean, has emphasized how the north is stirring up trouble in north-south relations by looking for an enemy in the south. According to Chater, Tunisian public opinion was first and foremost critical of the West's anti-immigration policies. The Tunisian people felt that they were being rejected. Any sense of a possible military threat posed by the north was definitely secondary.[59] Elsewhere, it has been pointed out that people in the southern Mediterranean are anxious that the cultural and religious rights of Maghreb communities living in Europe be respected and guaranteed. If this is perceived to not be the case, there could be a popular backlash in the south.[60] Therefore, there is clearly unease in the south with regard to the migration issue. Officials and commentators in the south are appealing for the north to treat the south on equal terms. Again, the south appears to be demanding more attention from the north.

One might question Chater's assertion that the south perceives the north as a military threat, albeit a secondary one. With the exception of the Libyan and possibly Sudanese governments, it is difficult to conceive how governing Arab elites in the southern Mediterranean are expecting an armed attack from the north in the foreseeable future. North African governments, though, are suspicious of the further development of the capabilities of European armed forces. They observe NATO and WEU exercises in the Mediterranean with unease. Arab officials are disturbed at talk of NATO or WEU conducting possible rescue and humanitarian operations in the southern Mediterranean.[61] These officials are less worried about a possible attack from the north. They are more concerned that the north may take advantage of its military supremacy to interfere in Arab internal politics. This study will later note how EUROFOR (European Rapid Operational Force) and EUROMARFOR (European Maritime Force), forces answerable to WEU, have been perceived by Arab governments.

There are grounds for cooperation in security issues between officials north and south of the Mediterranean. For example, the interior ministers from France, Italy, Spain, Portugal, Algeria, and Tunisia met in Tunis in January 1995 to discuss how to coordinate their activities in relation to the threat posed by radical, politicized Islam. In practice, in spite of the talk of a possible threat from the north, Arab ruling elites are still primarily concerned with internal threats. As noted previously, officials in the West, including NATO policy-makers, need to be careful not to emphasize the need for stability in North Africa and the Middle East at the expense of democracy. The West should also attempt to contact public opinion-makers as well as ruling officials. NATO in its Mediterranean Initiative is making efforts to build ties with such opinion-makers.

There is another problem that Western officials and NATO policy-makers have to address. Governing circles in Arab countries are appealing for the West to become more actively involved in tackling the social and economic problems of the southern Mediterranean. How can this be accomplished without the north being perceived by ruling elites or people in the south as attempting to interfere in internal Arab politics? Chater, for example, in effect argues that cooperation between states north and south of the Mediterranean should be based on preventive diplomacy and not crisis management, as the latter is more likely to lead to interference in internal affairs.[62] Following this line of argument, organizations such as NATO should aim to prevent crises from emerging rather than seek to contain or deal with crises that have already emerged. There would thus be a need for CBMS and CSBMS. In practice, though, NATO forces could still be involved in crisis-management operations in the Mediterranean—for instance, to rescue trapped Western civilians—if preventive diplomacy fails. It would seem, therefore, that NATO officials, in their discussions with Arab representatives, should emphasize more NATO's role in preventive diplomacy. At the same time, though, it is important for NATO to also spell out clearly what it means by "crisis management" in order to assure Arab countries that it is not some sort of euphemistic expression disguising an actual intention to interfere in the internal affairs of Arab states.

CONCLUSION

Obviously, the EU, rather than NATO, must assume a key role in attempting to alleviate economic, social, and political problems in North Africa and the Middle East. But these problems, if left unresolved, may lead to heightened tensions between states north and south of the Mediterranean. With its interest in developing CBMs and CSBMs in the Mediterranean, NATO does have an important part to perform in coordination with the EU and other bodies such as WEU, the OSCE, and the Mediterranean Forum in helping to enhance security in the Mediterranean area. NATO and other institutions are faced with the common task of ensuring that governments and peoples in the southern Mediterranean perceive the north not as a threat or as a disinterested neighbor but more as a partner willing to offer a hand in support.

At present the Mediterranean or Euro-Mediterranean area is one in which stable or cold peace prevails. Tensions and problems between states in the area—south-south as well as north-south—are still manageable. NATO and other bodies and institutions are seeking to bolster this peace and build confidence between governments and peoples in the area. In these circumstances NATO's Mediterranean Initiative is a worthwhile exercise. However, realistically speaking, given all the problems outlined in this and the previous chapter, it appears that there is little prospect in the foreseeable future for the formation of a security regime let alone a security community in the Mediterranean. The emphasis should be on gradually developing and strengthening a confidence-building process in the Mediterranean area. NATO's Mediterranean Initiative has a role to play in this process.

NOTES

[1] George Joffé, "Southern Attitudes towards an Integrated Mediterranean Region," *Mediterranean Politics* 2, 1 (summer 1997): 23.

[2] Andres Ortega, "Relations with the Maghreb," in *Maelstrom: The United States, Southern Europe and the Challenges of the Mediterranean,* John W. Holmes, ed. (Cambridge, Mass.: The World Peace Foundation, 1995), 34.

[3] Joffé, "Southern Attitudes towards an Integrated Mediterranean Region," 24–26.

[4] Figures are from the table in the appendix of Sarah Collinson, *Shore to Shore—The Politics of Migration in Euro-Maghreb Relations* (London, and Washington, D.C.: The Royal Institute of International Affairs and The Brookings Institution, 1996), 102–103.

⁵Nadji Safir, "The Question of Migration," in *Maelstrom*, 68.

⁶For more background details, see Collinson, *Shore to Shore*.

⁷Gil Loescher, "The European Community and Refugees," *International Affairs* 65, 4 (autumn 1989): 621.

⁸Note, though, that the Spanish authorities are anxious not to impose too many restrictions on immigration through fear that this could provoke an upsurge in anti-Spanish sentiment in Morocco, and in the enclaves of Ceuta and Melilla.

⁹George Joffé, "The European Union and the Maghreb," in *Mediterranean Politics, Vol.1*, Richard Gillespie, ed. (London: Pinter, 1994), 29.

¹⁰NAA, Political Committee, Sub-Committee on the Southern Region, *Cooperation and Security in the Mediterranean*, AL 223 PC/SR (94) 2 by Mr. Rodrigo de Rato (Spain), Rapporteur (Brussels: International Secretariat, November 1994), 8.

¹¹Tim Niblock, "North-South Economic Relations in the Mediterranean," in *Security Challenges in the Mediterranean Region*, Roberto Aliboni, George Joffé, and Tim Niblock, eds. (London, and Portland, Oreg.: Frank Cass, 1996), 126–27.

¹²Jonas Widgren, "International Migration and Regional Stability," *International Affairs* 66, 4 (October 1990): 755.

¹³Roberto Aliboni, *European Security across the Mediterranean* (Paris: Chaillot Papers 2, Institute for Security Studies, WEU, 1991), 21.

¹⁴Widgren, "International Migration and Regional Stability," 754.

¹⁵Joffé, "The European Union and the Maghreb," 30; and Niblock, "North-South Economic Relations," 121.

¹⁶Safir, "The Question of Migration," 72–73.

¹⁷Joffé, "Southern Attitudes towards an Integrated Mediterranean Region," 22–23.

¹⁸Habib Ben Yahia, "Security and Stability in the Mediterranean: Regional and International Challenges," *Mediterranean Quarterly* 4, 1 (winter 1993): 6.

¹⁹Safir, "The Question of Migration," 77–78.

²⁰Ortega, "Relations with the Maghreb," 49.

²¹Collinson, *Shore to Shore*, 58–60.

²²Marjorie Lister, *The European Union and the South: Relations with Developing Countries* (London and New York: Routledge, 1997), 101.

²³NAA, Civilian Affairs Committee, Sub-Committee on the Mediterranean Basin, *Report—Seminar on the Maghreb, Paris, 9–10 June 1994*, AL 155 CC/MB (94) 3 (Brussels: International Secretariat, July 1994), 3.

²⁴Widgren, "International Migration and Regional Stability," 758–59.

²⁵Safir, "The Question of Migration," 64, 74.

²⁶Collinson, *Shore to Shore*, 46.

[27] Michael Willis, "The Islamist Movements of North Africa," in *Security Challenges in the Mediterranean Region*, 21.
[28] Collinson, *Shore to Shore*, 47–48.
[29] Ibid., 62.
[30] Simon Serfaty, "Algeria Unhinged: What Next? Who Cares? Who Leads?" *Survival* 38 4 (winter 1996–97): 140–41.
[31] Cassandra, "The Impending Crisis in Egypt," *Middle East Journal* 49, 1 (winter 1995): 11–14.
[32] Marat Terterov, "Egypt: The Islamist Challenge," *Mediterranean Politics* 1, 2 (autumn 1996): 244.
[33] "A Revolution to End the Revolution," *The Economist*, October 25, 1997.
[34] Rodney Wilson, "Economic Prospects for Iraq and Its Neighbours in the Aftermath of the Gulf War," in *The Middle East in the New World Order*, 2d ed., Haifaa A. Jawad, ed. (London and Basingstoke: Macmillan, 1997), 191–93.
[35] Anthony G Pazzanita, "Political Transition in Mauritania: Problems and Prospects," *Middle East Journal* 53, 1 (winter 1999): 56, 58.
[36] For more details on recent developments in the Moroccan economy, see George Joffé, "Elections and Reform in Morocco," in *Mediterranean Politics, Vol.1*, 221–24; and Josep M. Jordan Galduf, "Spanish-Moroccan Economic Relations," *Mediterranean Politics* 2, 1 (summer 1997): 52–53.
[37] Nicole Grimaud, "Tunisia: Between Control and Liberalization," *Mediterranean Politics* 1, 1 (summer 1996): 101–102.
[38] Emma C. Murphy, "Ten Years on—Ben Ali's Tunisia," *Mediterranean Politics* 2, 3 (winter 1997): 115–17.
[39] Rosemary Hollis, "Europe and the Middle East: Power by Stealth?" *International Affairs* 73, 1 (January 1997): 16.
[40] Niblock, "North-South Economic Relations in the Mediterranean," 118–19.
[41] Hollis, "Europe and the Middle East," 16.
[42] Joffé, "The European Union and the Maghreb," 39–40.
[43] Lister, *The European Union and the South*, 93–95.
[44] "Arab Trade: With Whom?" *The Economist*, October 10, 1998.
[45] Christophe Carle, "France, the Mediterranean, and Southern European Security," in *Southern European Security in the 1990s*, Roberto Aliboni, ed. (London and New York: Pinter, 1992), 47.
[46] *The Military Balance 1994–95* (London: Brassey's, for the International Institute for Strategic Studies, 1994), 153.
[47] Keith B. Payne, "Post–Cold War Deterrence and Missile Defence," *Orbis* 39, 2 (spring 1995): 203–04.
[48] Martin Navias, *Going Ballistic—The Build-Up of Missiles in the Middle East* (London: Brassey's, 1993), 29.

⁴⁹Ibid., 21.

⁵⁰*Assembly of Western European Union, Proceedings, 42nd Session, December 1996, I Assembly Documents* (Paris: WEU), Doc. 1543, November 4, 1996, "Security in the Mediterranean Region." Report submitted on behalf of the Political Committee by Mr. de Lipkowski, Rapporteur, 172.

⁵¹Fernanda Faria, "Security Policies and Defence Priorities," in *Security in Northern Africa: Ambiguity and Reality,* Fernanda Faria and Alvaro Vasconcelos (Paris: Chaillot Papers 25, Institute for Security Studies, WEU, 1996), n.54, 30.

⁵²NAA, Civilian Affairs Committee, Sub-Committee on the Mediterranean Basin, *Report—Frameworks for Cooperation in the Mediterranean,* AM 259 CC/MB (95) 7 by Mr. Pedro Moya (Spain), Rapporteur (Brussels: International Secretariat, October 1995), 13.

⁵³Faria, "Security Policies and Defence Priorities," 43.

⁵⁴Ibid., 4, 5, 26.

⁵⁵Collinson, *Shore to Shore,* 51–53.

⁵⁶Aliboni, "Introduction," in *Security Challenges in the Mediterranean Region,* x.

⁵⁷Ben Yahia, "Security and Stability in the Mediterranean," 2.

⁵⁸Carle, "France, the Mediterranean, and Southern European Security," 48.

⁵⁹Khalifa Chater, "Mediterranean Security: The Tunisian Viewpoint," in *Security Challenges in the Mediterranean Region,* 74–75.

⁶⁰Alvaro Vasconcelos, "General Framework and Concepts," in *Security in Northern Africa,* 19.

⁶¹Monica Wohlfeld, "A Survey of Strategic Interests of the Countries of the European Security Space," in *The European Security Space,* Guido Lenzi and Laurence Martin, eds. (Paris: Institute for Security Studies, WEU, 1996), 114; and Margaret Blunden, "Insecurity on Europe's Southern Flank," *Survival* 36, 2 (summer 1994): 138–39.

⁶²Chater, "Mediterranean Security," 77.

CHAPTER 6
The Immediate Background to NATO's Mediterranean Initiative

INTRODUCTION

As previously discussed, throughout most of the Cold War period NATO's concerns in the Mediterranean as a whole focused on Soviet activities. In the eastern Mediterranean the United States, in particular, was eager to lend assistance and support to Israel. By the 1980s, though, NATO member states in the Southern Region were becoming increasingly apprehensive about developments to their immediate south and east. These NATO states had become aware that economic, social, and political problems in North Africa could have wider negative consequences for the security and stability of the Mediterranean. In the same period, Turkey was forced to focus more on its relations with Iran, Iraq, and Syria as the Kurdish problem attracted more attention. The possible threat posed by ballistic missiles and WMD in the hands of Arab governments or Iran was being discussed in NATO circles. A debate had begun between various NATO member states concerning the out-of-area issue, although NATO as an organization was not perceived as equipped to handle problems out-of-area.

In the post–Cold War era, with NATO as an organization potentially more free to operate out-of-area given the demise of the Soviet threat, there were three significant developments that had a major impact on NATO-Mediterranean relations and that dramatically changed the whole context of the out-of-area debate. The Iraqi invasion of Kuwait, followed by Operation Desert Storm, must first be considered. In practice, it was not a NATO- but US-led international coalition force that ousted Iraqi forces from Kuwait. However, NATO member states played a key role in the coalition force, and NATO as an organization was involved in naval patrols in the Mediterranean and in taking precautionary action to deter an Iraqi attack on Turkey. The crisis in former Yugoslavia and in Bosnia, in particular, also has to be taken into account. NATO was clearly involved in an out-of-area operation in Bosnia, although the Atlantic Alliance failed to act effectively for a long period largely because of the lack of political will and the difficulties involved in establishing a proper chain of command with the UN. In spring 1999 NATO authorities were determined to adopt a much more vigorous stance with regard to Kosovo. Finally, positive events in the Middle East, leading to the start of a peace process in the region, also must be noted. This was in part a consequence of the end of the Cold War, as the Soviet Union was no longer willing to sponsor radical groups in the Middle East. Agreements between Israelis and Palestinians, a peace treaty signed by Israel and Jordan, and a series of multilateral talks involving more states in the region and outside interested parties raised hopes that a comprehensive Middle East peace settlement was attainable in the foreseeable future.

With the end of the Cold War, security issues concerning the Middle East and the Gulf began to have an increasing impact on the security and stability of states in the southern Mediterranean as a whole. Arab governments no longer had the option of courting two competing superpowers to obtain economic, political, or military support. The end of East-West rivalry allowed more attention to be focused on north-south issues in the Mediterranean or Euro-Mediterranean area. Problems in Arab-Israeli relations could be directly addressed without the complicating backdrop of Soviet-American competition in the Middle East. However, in these new circumstances, a failure to conclude satisfactorily the Middle East peace process could have more serious destabilizing consequences for north-south relations in the Mediterranean as a whole, given that most Arab governments and peoples in the area tend to associate Israel with the north (or West).

THE SIGNIFICANCE OF DEVELOPMENTS IN THE GULF, THE BALKANS, AND THE MIDDLE EAST

The impact of the Iraqi invasion of Kuwait on the Arab world with the spontaneous popular demonstrations in support of Saddam Hussein throughout North Africa and the Middle East has been previously noted. Spain and Portugal were alarmed on learning that Iraq could have deployed SCUD missiles in Mauritania. Such missiles would have been within striking distance of the Canary Islands and Madeira.[1] Iraq, after all, had fired SCUD missiles at Israel and Saudi Arabia and although these missiles were only tipped with conventional warheads the SCUD attacks had caused widespread disruption. Units of NATO's Allied Command Europe–Air Mobile Force were dispatched to Turkey in January 1991 to forestall a possible Iraqi assault. The Air Mobile Force had never previously been deployed. The "threat from the south" in the form of ballistic missiles and possibly WMD was forced on to the Atlantic Alliance's agenda. Following the Iraqi invasion, NATO naval forces and aircraft were alerted to protect air and sea lines of communication in the Mediterranean. Minesweepers were also dispatched in the Mediterranean as it was feared that Iraq and perhaps Libya might attempt to hinder Western access to oil supplies.

As the situation deteriorated in the Gulf in the weeks before the launching of Operation Desert Storm, NATO Secretary-General Manfred Wörner in a speech at the end of November 1990 referred to the emergence of "an arc of tension from the Maghreb to the Middle East." Wörner spoke of the problem of WMD in this area. He noted that tensions were aggravated by the presence of dictators like Saddam Hussein. Wörner also referred to such problems as rapid population growth, resource conflicts, migration, economic underdevelopment, and the spread of religious fundamentalism and terrorism as other sources of tension.[2]

The Iraqi invasion of Kuwait had forced NATO to focus more attention on out-of-area issues within an "arc of tension" that stretched across the Mediterranean to the Gulf.

The crisis in Bosnia had repercussions in the Arab and Muslim worlds. Comparisons were drawn between the swiftness and effectiveness of the international coalition in its defeat of Iraqi forces in Operation Desert Storm and the reluctance of NATO member states to use decisive firepower against the Bosnian Serbs. It was pointed out that in the Bosnian case there was no oil and the principal victims were Muslims not Christians. One should note, though, that with reference to Bosnia, there was a considerable difference in the way Muslim governments and their peoples criticized the perceived inaction of the West. Officials in Muslim countries complained of the lack of coordination between the European governments and the absence of a political will to become more militarily engaged. Whereas the Islamic media, and militant Islamic groups especially, focused more on the argument that there was a general European hostility to Islam and declared that there was a conspiracy against the Muslim world. These Islamic groups were pushing this line largely in order to boost their popularity at home. They were helped in their cause by various references among Western scholars and commentators to "a clash of civilizations," "the green peril," and the "Islamic threat."[3]

On a more positive note for NATO, units from Egypt, Jordan, and Morocco later participated with NATO troops in Bosnia in the multilateral peacekeeping forces known as IFOR and SFOR. This was a good example of cooperation between NATO and the armed forces of three important non-NATO "Mediterranean" countries. The Egyptian foreign minister even told WEU Rapporteur de Lipkowski that the participation of Arab troops in Bosnia could serve as a model for Euro-Mediterranean cooperation in crisis-management situations in other regions and might pave the way for cooperation in other areas.[4]

Units from Egypt, Jordan, Morocco and, Tunisia (and also from Algeria) have been involved in police work and taken part in military observation missions in Bosnia, Croatia, and Macedonia. Personnel from Jordan and Tunisia (and Algeria) have also helped in humanitarian assistance programs in Bosnia.[5]

Certainly, Western governments could have acted more decisively in Bosnia before NATO finally bombarded Bosnian Serb positions in late summer 1995. Nevertheless, NATO as an organization had been involved in Bosnia in a number of out-of-area roles. In line with a decision taken by NATO ambassadors in September 1992, individual allies provided troops to protect and escort humanitarian aid convoys in the region under UN command. Following a decision taken by the NAC in Brussels in December 1992, whereby NATO proposed to support peacekeeping operations under the mandate of the UN Security Council, NATO units and resources were used to support the headquarters of the UN Protection Forces (UNPROFOR) in Bosnia. Together with a flotilla under WEU authority, NATO vessels enforced a UN maritime embargo on the former Yugoslavia. NATO aircraft enforced a UN ban on unauthorized flights over Bosnia and then were employed in limited air strikes in coordination with the UN to maintain military exclusion zones and protect UNPROFOR units. NATO units also helped to enforce a cease-fire around Sarajevo and in central Bosnia. In former Yugoslavia, therefore, NATO forces participated in various out-of-area missions—in peace-making, sanctions-enforcing, and cease-fire-enforcement tasks—before commencing peace-building operations as a part of IFOR and then SFOR.

In spring 1999 NATO launched an air bombardment against Serbia after the escalation of the crisis in Kosovo. In this instance, neither Arab governments nor Arab public opinion could accuse NATO of not taking a firm stance against Belgrade. However, pressure from the international community soon mounted for NATO to wage a land campaign against Serbian forces.

Positive developments in the Middle East had potential beneficial consequences for all of the Mediterranean. But, more so than their governments, Arab peoples remained suspicious of Israeli and Western intentions. The launching of the Middle East peace process after the Madrid meeting of October 1991 has been discussed earlier. In the multilateral framework of talks that were initiated in January 1992, five working groups were created to discuss economic cooperation, the environment, water resources, refugees, and arms control throughout the area. The purpose was to address so-called region-wide problems at a "regional" level in the hope that the multilateral talks could facilitate progress at the bilateral level. The multilateral talks focused on the possibilities of functional cooperation.[6] North African states and also the international community were encouraged to participate in the multilateral working groups. These groups concentrated on activities that were in effect CBMs. Arab governments started to cooperate with Israel in addressing various "regional" problems within the working groups. However, Lebanon and Syria refused to be a part of the multilateral discussions contending that a political settlement at the bilateral level should be reached first.

The largest of the working groups, and arguably the one with the most potential, was the Working Group on Regional Economic Development (REDWG). It was decided that a permanent monitoring committee of the parties from the area would be set up to oversee REDWG, establish a list of priorities for it, and be responsible for the EU's Copenhagen Action Plan, which aimed to boost "regional" economic cooperation by using REDWG as a vehicle. The EU thus began to help coordinate the work of REDWG. Within the parameters of the Action Plan, REDWG sought to promote "regional" trade and investment, develop "regional" economic infrastructures, and encourage the free movement of goods, services, capital, and information among the "regional" participants.[7]

An important beneficial spin-off of the activities of REDWG was the convening of the first MENA economic conference in Casablanca in October–November 1994. This was an initiative of the Moroccan government that was eagerly supported by Israel, in particular. Over one thousand leading politicians and businessmen from sixty-one countries attended the Casablanca meeting, but Syria and Lebanon were conspicuous by their absence. There was talk of a Casablanca process. It was agreed to hold further MENA economic conferences on a regular basis. Private entrepreneurs were beginning to become interested in promoting stability in the area by seeking to boost economic development. At Casablanca it was proposed that a Middle East and North Africa Development Bank, a Regional Chamber of Commerce and Business, and a Regional Tourism Board should be set up. A permanent secretariat responsible for organizing MENA meetings was later established in Casablanca. Further MENA economic conferences were held in Amman (1995), Cairo (1996), and in Doha, Qatar (1997). Unfortunately, as a result of problems in the Middle East peace process, no MENA economic conference was convened in 1998. The activities of all of the multilateral working groups had already been suspended.

One potentially very important group was the previously mentioned ACRS that was concerned with arms control and "regional" security. ACRS spearheaded the first concerted attempt to tackle issues pertaining to security in the Middle East involving Israel and a number of Arab parties. In Cairo in February 1994, ACRS produced the draft text of a "Declaration of Principles and Statements of Interest on Arms Control and Regional Security," which was modeled on and inspired by the Helsinki Final Act of 1975. This declaration referred to core principles for the peaceful resolution of conflict. It stressed the value of a step-by-step approach, emphasized the need for decision-making by consensus, spoke of the role of CSBMs, underlined the importance of conventional arms control, and advocated the establishment of a zone free of all WMD in the Middle East.[8] The Israel-Jordan peace treaty of 1994 specifically mentioned the valuable role of ACRS in promoting "regional peace."

In effect, ACRS was working on a series of first generation CBMs. Attempting to model its activities on the CSCE experience, ACRS began exploring the possibilities for the prior notification of certain military exchanges, the exchange of military officers, the exchange of unclassified publications and education and training manuals, voluntary invitations to visit defense installations, the creation of a communications network, and joint search and rescue operations at sea. An ACRS Communications Network based in Cairo was formally established in spring 1995. Approval was also given to set up a series of "regional" security centers—conflict prevention centers—in Amman, Tunis, and Qatar. In Amman in September 1995, a group of experts had agreed on what tasks these centers should carry out with regard to seminars, education, and training.[9]

There were shortcomings in ACRS. The "rogue states" Iran, Iraq, and Libya were not invited to participate in the group. Syria and Lebanon were invited but refused to attend. Thus, ACRS was not able to deal comprehensively with security issues concerning the Mediterranean and the Middle East. The security of Israel, in particular, was to a great extent related to the policies of these absentee states. A major weakness of ACRS was that it was an "unabashed creature of the peace process," which thus set limits on its membership and on its vision. The focus was limited to Arab-Israeli concerns and excluded Arab-Arab problems, questions of internal stability, and the role of outside states in the area. In reality, ACRS was not properly equipped to handle these issues, and, arguably, the United States had no desire to extend the group's remit.[10]

In the course of 1994 the original Declaration of Principles was watered down. Saudi Arabia was opposed to the term "principles," which, they felt, implied that issues such as pluralism and human rights might be raised as they had been in Helsinki. By the end of 1994 the Declaration had been further downgraded and had become merely a "Statement on Arms Control and Regional Security."[11] By this time the activities of ACRS were beginning to grind to a halt as a result of a dispute between Egypt and Israel. The Egyptian delegates demanded that the issue of WMD, and, in particular, the question of Israel and its ownership of nuclear weaponry, should be tackled before other CBMs could be seriously discussed. The Israeli position was that attention should first be given to CBMs in general. At one time Egypt threatened to withhold support for the indefinite extension of the NPT until Israel had become a signatory and was prepared to open its nuclear facilities to international inspection. Under US pressure Egypt was forced to back down, but then Egyptian officials linked the nuclear issue and Israel to the whole ACRS process. With Israel not prepared to change its stance, by the end of 1995 the Arab states refused to cooperate with ACRS, and the activities of the working group were suspended. Work was halted on the development of the ACRS Communications Network.

In general, however, in spite of the problems that ACRS was encountering by early 1995, there had been positive developments in the Middle East in the period 1991–spring 1995. As an alliance, obviously, NATO was not directly responsible for this turn of events, but NATO was able to exploit the changing circumstances to its advantage and eventually initiate official contacts with certain non-NATO Mediterranean countries. This was because positive developments in the Middle East improved the prospects for more stability and enhanced security in the

Mediterranean as a whole. The difficulties in ACRS and the failure to hold further MENA economic conferences notwithstanding, the potential beneficial economic spin-offs for the area of the work of REDWG in particular, and also of the other multilateral working groups, if they can resume their work, needs to be emphasized again.

The Gulf War, Bosnia, and the beginning of the Middle East peace process all had an impact on the out-of-area debate in Atlantic Alliance circles. The distinction between in-area and out-of-area was becoming increasingly blurred. In a speech at Annapolis in June 1992, Secretary-General Wörner noted that the concept out-of-area made sense in the Cold War era when any military operations by NATO beyond its territorial jurisdiction might escalate into an East-West confrontation. But "in a new security environment traditional notions like 'in' and 'out-of-area' are increasingly losing their significance."[12] Wörner, here, had implied that in a post–Cold War environment it would be less dangerous for NATO to venture out-of-area. This would soon prove to be the case when NATO became involved in what amounted to a crisis management operation out-of-area in Bosnia. The in-area/out-of-area distinction was also losing relevance because of NATO's involvement with central and east European states through the activities of the North Atlantic Cooperation Council (NACC) and the later Partnership for Peace (PfP) programs. Were these examples of a new Atlantic Alliance out-of-area policy, or had the term "out-of-area" become an outdated and redundant one?

Immediately after Operation Desert Storm, NATO foreign ministers meeting in Copenhagen in June 1991 declared that following the Gulf War experience the Atlantic Alliance "must be prepared to address other unpredictable developments that are beyond the focus of traditional alliance concerns, but that can have direct implications for our security."[13] It was almost as if the foreign ministers were aware of the impending significance for NATO of another "development," which was already unfolding at the time in Yugoslavia.

In the words of a report of the NAA published in 1993, the debate was no longer about whether NATO should operate in territory traditionally regarded as out-of-area, but, rather, "over how out-of-area policies should be pursued." The issue of territory was also losing significance. The same report noted: "In future, NATO will be called upon to defend Alliance **interests** [original in bold] as much as defend the territory of its member states." This was followed by references that stressed that NATO should not allow the disruption of energy supplies and should prevent states from acquiring WMD.[14] The emphasis was thus on defending

NATO's "interests" wherever these interests might be. By this time NATO policy-makers were repeatedly underlining the need for the Atlantic Alliance to become more concerned and more prepared to be involved in preventive diplomacy and crisis management in territory that was traditionally regarded as out-of-area.

In line with a trend that had been increasingly evident since the early 1980s, NATO decision-makers also referred more to the nonmilitary aspects of security. The Atlantic Alliance's London Declaration of July 1990 noted: "We reaffirm that security and stability do not lie solely in the military dimension and we intend to enhance the political component of our Alliance as provided for by Article Two of our Treaty."[15]

In November 1991 at the Rome Summit the new Strategic Concept of NATO referred to the importance of the "broad concept of security," which included political, economic, social, and environmental elements in addition to the military and defense dimensions.[16]

Clearly, NATO was adjusting to the end of the Cold War and the disappearance of the massive Soviet threat. In place of this threat, there was a growing apprehension about "risks" emerging out-of-area that might be economic, social, environmental, political, and also military in nature. The military aspect of security could not be excluded given recent events in the Gulf and NATO's continued concern about the proliferation of WMD. The enlarged concept of security extensively discussed at Rome would give NATO more freedom of maneuver to engage in dialogue with other states.

Another indication of the continued importance of the military and defense dimension of security for NATO was evident in the reorganization of the Atlantic Alliance's force structures in the Mediterranean. Reorganization was required in order to adjust to the new post–Cold War environment. The experience of the Gulf War and then the crisis in the Balkans accelerated the process of restructuring NATO's forces in the Mediterranean. A priority was for NATO to possess forces capable of reacting swiftly to prevent or contain crises. Immediate and rapid reaction forces with ground, maritime, and air components were formed. Thus, on April 30, 1992, the Standing Naval Force Mediterranean (STANAVFORMED) was formally activated. In contrast to its predecessor, NAVOCFORMED, which was only periodically activated, STANAVFORMED was intended to be a permanent immediate-reaction naval force. This new force soon served in a crisis management role when sanctions were applied against Yugoslavia. AFSOUTH's contribution to the new Allied Rapid Reaction Corps was the formation of a Multinational Division South. It was also decided to establish a combined amphibious force and a multinational mine-countermeasures force to serve in the Mediterranean.[17] Meanwhile, Spain was preparing its own rapid-reaction force. By 1993 France, Spain, and Italy—the first two at the time not a part of NATO's integrated military structure—were planning to coordinate their armed forces in the Mediterranean area. The end product of their initiative would be the announcement of the formation of EUROFOR and EUROMARFOR in May 1995 as forces answerable to WEU.

NATO also focused more on the problem of WMD, a classic hard-security issue. The Alliance's Strategic Concept of November 1991 had referred to the problem of WMD. In June 1994 the NAC adopted the Alliance Policy Framework on the Proliferation of Weapons of Mass Destruction.[18] A Senior Political-Military Group was set up within the Atlantic Alliance to work with other international bodies to attempt to prevent proliferation or reverse it by diplomatic means. A Senior Defense Group was also established on the recommendation of the Alliance Policy Framework to work on how to protect NATO territory and forces in the event of a military attack. Both groups could pool their efforts in a Joint Committee on Proliferation. Various air defense programs and sea-based systems were soon being developed. In December 1994 NATO foreign ministers and defense ministers endorsed the work of the Senior Defense Group, which had completed a risk assessment. In this assessment (the first of its kind produced by NATO) that considered likely technological advancements to the year 2010, the main conclusion was that the proliferation of nuclear, biological, and chemical weapons could pose a direct threat to the Atlantic Alliance and must be taken into account in defense planning. Further work was carried out assessing proliferation problems. Specific recommendations were made to improve the military capabilities of NATO to contend with the possible proliferation of WMD.[19]

It will be seen that this concern in NATO circles over arms proliferation, with its implication that a threat from the south was real, was deliberately not directly addressed by NATO officials in their dialogue with non-NATO Mediterranean countries. The proliferation issue is a sensitive one, especially when Arab public opinion, in general, feels that it is the north that poses a threat to the south and not vice versa. NATO policymakers are careful not to overstate the risk of proliferation of WMD. The risk could then become a self-fulfilling prophecy.[20] If NATO were to spend considerable sums of money to create a theater nuclear-defense system, this could be regarded by the Arabs in North Africa and in the Middle East as hostile and could well exacerbate north-south tensions in the Mediterranean. In practice, though, for the foreseeable future, it seems that Arab states are more likely to deploy their missiles and aircraft against other Arab states or Israel than against NATO. However, NATO forces may have to confront WMD in the field if they are ever to become involved in crisis-management operations in the southern or eastern Mediterranean. One such operation, for example, may involve the need to rescue trapped Western civilians. It is also important to note that those states that potentially pose more of a direct threat to NATO because of their interest in acquiring WMD are not states that figure in NATO's Mediterranean Initiative. Taking the Mediterranean and the Gulf as a whole, it is the rogue states of Iran, Iraq, Libya, and Syria—also referred to by some commentators as the "Club Mad" grouping of states—that may threaten NATO member states with their WMD in the future.

FAILED TRANS-MEDITERRANEAN COOPERATION INITIATIVES

In the late 1980s and early 1990s France, Italy, and Spain had become interested in promoting trans-Mediterranean cooperation initiatives. This was in part because of their concern regarding the possible spillover to southern Europe of economic, political, and social problems in North Africa in such forms as migration, the spread of religious radicalism, and terrorism. Southern Europe was also within range of a theoretical air or ballistic missile strike from the south. Spain and Italy, in particular, were also to an extent dependent on energy supplies from the Maghreb. Furthermore, states in NATO's Southern Region were anxious not to become marginalized within the Atlantic Alliance as the United States, Canada, and north European allies tended to give more attention to central and eastern Europe and the states of the former Soviet Union. Although each had their own separate interests and perceptions, France, Italy, Spain, and also Portugal—the latter three the so-called Club Med grouping of states within NATO—held a common belief that the problems of North Africa should not be overlooked.[21]

One commentator writing in the early 1990s suggested that if the Atlantic Alliance as a whole did not give enough attention to the southern Mediterranean, there was a possibility of segmentation within NATO. A specialized southern group of allies could emerge prepared to deal with any threat from the south. This could even lead to the renationalization of the security and foreign policies of the allies concerned. Tensions could ensue between the southern Europeans and the United States.[22] Arguably, the beginning of a NATO-Mediterranean dialogue helped to forestall the possible de facto fragmentation of the Atlantic Alliance along a north-south axis.

Although France was traditionally interested in the Mediterranean as a whole, in recent years French officials have tended to concentrate more on the western Mediterranean as the United States consolidated its presence in the Middle East and in the Gulf. Because of their special and at times problematic relationship with the Atlantic Alliance, French policymakers were initially inclined to encourage the EU and WEU to become more active in the Mediterranean than in NATO. However, the limitations of the EU and WEU in the Balkans must have compelled French officials to reassess their position. Nevertheless, with the end of the Cold War, in addition to seeking to strengthen its bilateral ties with the Maghreb states, France attempted to sponsor various multilateral initiatives in the western Mediterranean.

In addition to Spain's traditional ties with Morocco, and Madrid's apprehension over the future of the enclaves of Ceuta and Melilla, the Spanish government was also eager to continue to import gas from Algeria, and gas and oil from Libya. A more confident and stable Spain was emerging after the decades of repression under the Franco regime. Spanish officials were seeking to play a more heightened role in the Mediterranean. Italy's newfound interest in the Mediterranean was further highlighted with the emergence on the political scene of Gianni de Michelis as foreign minister. The new foreign minister was particularly keen on Italy promoting multilateral initiatives in the Mediterranean and also in central and eastern Europe. In November 1989 he announced the formation of the Quadrangolare which consisted of Italy, Yugoslavia, Austria, and Hungary. This grouping of states would eventually form the core of the larger but much less effective Central European Initiative.

By the late 1980s Portugal was also shifting more of its attention to the Mediterranean, as the migration issue came on the agenda. Portugal had concluded an agreement on military and political cooperation with Morocco. The Portuguese were importing oil products from Algeria and Libya, and gas from Algeria.[23]

Turkey and Greece were less interested in trans-Mediterranean cooperation initiatives, as they were too distracted with their own problems in the Aegean and over Cyprus, and were anxiously following events in the Balkans. Turkey was also closely monitoring developments in Central Asia and in the Transcaucasus, and maintained a close watch on what was happening in the Kurdish-populated regions of northern Iraq in the immediate aftermath of the Gulf War. Turkey, however, did sponsor a multilateral regional initiative of its own in the form of the Black Sea Economic Cooperation.

As previously noted, by 1989 a regional grouping of states, known as the AMU, had formed in the southern Mediterranean. The formation of the AMU would facilitate the promotion of trans-Mediterranean cooperation initiatives as the southern European states had a recognized interstate entity in North Africa with whom they could negotiate. Algeria, Morocco, and Tunisia in particular would play a leading role in various trans-Mediterranean groupings and would also be included in various north-south dialogues in the Mediterranean launched by international bodies and organizations. However, the military intervention in Algeria in 1992 would create problems. Likewise, Libya's membership in AMU would become troublesome after UN sanctions were imposed on Libya after the Lockerbie air disaster. But Mauritania's membership in the AMU would enable this desert state to be included in various Mediterranean initiatives even though Mauritania was certainly not a Mediterranean state in terms of geography.

Even before de Michelis became foreign minister, Italy attempted to foster a sense of Mediterranean solidarity through the so-called Group of Contact. This was an attempt at the end of 1986 to make use of the model of the successful Contadora Group in Central America. The objective was to set up a regional "support group" of countries that could hopefully deal with any future crises in the Mediterranean area. This rather vague initiative was never taken up.

The attempt to launch a CSCM was a much more serious proposition. Italy and Spain sponsored this trans-Mediterranean project. It was first officially proposed at a CSCE conference held at Palma de Mallorca in September and October 1990. This conference had been convened in order to discuss the ecological system of the Mediterranean. Until this time the CSCE had not given much attention to the Mediterranean. The CSCE had instead primarily worked on human rights issues and security problems within an East-West context. Apparently, in 1973 Maltese Prime Minister Dom Mintoff had insisted that the CSCE should consider having a Mediterranean dimension.[24] Evidently, Malta, Cyprus, Spain, Italy, and Yugoslavia had demanded the assertion of the clauses referring to the Mediterranean in the 1975 Helsinki Final Act.[25]

A section on questions relating to security and cooperation in the Mediterranean was therefore included in the Final Act. It was noted that security problems in the Mediterranean and in Europe were interlinked. Reference was made to the need to develop a relationship with the "nonparticipating Mediterranean states." The possibilities of cooperation in issues relating to economics and the environment were underlined. The Final Act also noted that there was already a "dialogue" between the participating states and the nonparticipating Mediterranean states, and recommended that this dialogue and other contacts be maintained and amplified. It was suggested that all states in the Mediterranean be included in order to enhance peace, security, and cooperation in the region.[26]

It was not clear when this CSCE "dialogue" with the Mediterranean had commenced. In practice, this ostensible dialogue achieved little. The Helsinki Final Act of 1975 was full of ambiguous references to the Mediterranean. For example, what was meant by the proposal that all the states in the so-called Mediterranean region should be included in a dialogue? One is confronted with the question of where the boundaries of the Mediterranean lie. And as one analyst has noted, the title "nonparticipating Mediterranean states" was "an awkward status inferior to that of observer."[27] The nonparticipating Mediterranean states that attended the 1975 Helsinki Summit, and the CSCE conferences that specialized on the Mediterranean (Valletta, February 1979; Venice, 1984; and Palma de Mallorca), were Algeria, Egypt, Israel, Lebanon, Libya, Morocco, Syria, and Tunisia.

At the CSCE Venice Conference in 1984, which was titled "Economic, Scientific, and Cultural Cooperation in the Mediterranean Basin," then Italian Foreign Minister Guilio Andreotti had raised the issue of promoting a CSCM.[28] In December 1989 at the Paris ministerial meeting of what had by then become the largely moribund Euro-Arab Dialogue, De Michelis argued that "the time had now come to extend the spirit and rationale of Helsinki to the Mediterranean and the Middle East."[29] This was picked up by Spanish Foreign Minister Francisco Fernandez Ordonez who in February 1990 at the Ottawa "Open Skies" Conference called for a CSCM.[30] The Greek Foreign Minister Antonis Samaras supported the idea of a CSCM in a declaration to the UN General Assembly on September 27, 1990.[31] Egypt also apparently actively backed these proposals while France had a "non-negative" attitude.[32] As cosponsors of the CSCM, Italy and Spain circulated an "Italian-Spanish Nonpaper on CSCM" at Palma de Mallorca. This led to a Joint Document being issued by France, Italy, Portugal, and Spain at the meeting. It seems that France and Portugal, with some reluctance, had been brought on board.

The Joint Document called for the convening of a CSCM to define a "set of generally accepted rules and principles." The intention was to set in motion a process that would lead to the adoption of a "catalogue of principles" in the form of a "Mediterranean Act" that would "set the rules of behavior and coexistence in the area." The "Mediterranean region" was referred to in a "broad sense, including the Middle East and the Gulf" and also incorporating those states having an interest in the Mediterranean. In this context, the EC, the United States, the Soviet Union, and also the Palestinians were mentioned. Taking the Helsinki Final Act of 1975 as a model, the Joint Document listed three baskets of issues that should be covered—security, economic cooperation, and a human dimension. The third basket would deal with "dialogue, tolerance, and understanding among societies, civilizations, and beliefs" and referred to the need to work out a "common approach to human rights both in theory and practice." As soon as a "sufficient measure of consensus" had been obtained, a Preparatory Committee could be convened in order to inaugurate a formal phase in the CSCM process.[33]

The official Final Report of the meeting in Palma de Mallorca was much more cautious and noncommittal in its wording. "When circumstances allowed," a meeting "outside" the CSCE could take place that should be inspired by the experiences in the CSCE process and that "could discuss a set of generally accepted rules and principles in the fields of stability, cooperation, and the human dimension in the Mediterranean." The Final Report spoke of the need to introduce far-reaching economic and political reforms in the Mediterranean. A multiparty system should be established and the human rights of all citizens should be respected.[34] There was an implication that these reforms were not likely in the short term and that "circumstances" would not allow a meeting "outside" the CSCE to take place in the foreseeable future. One may contend that no such meeting was convened until the Barcelona Conference assembled in autumn 1995 under the auspices of the EU. It appears that a failure to secure consensus on various issues at Palma de Mallorca led to a considerable watering down of the original Joint Document.

France and Portugal were more interested in devising cooperative schemes that would concentrate on the western Mediterranean and not cover the Middle East and the Gulf. Portugal thus supported French efforts to promote a "5+4" and then "5+5" arrangement in the western Mediterranean.[35] To French officials, the CSCM was at best a long-term prospect. Conflicts in the eastern Mediterranean first required resolving.³⁶ At Palma de Mallorca there was also the problem of defining the so-called Mediterranean region. Italy favored including all states from Mauritania in the west to Pakistan in the east. The United States, Britain, and Germany were least enthusiastic about the CSCM project. The United States was afraid that the naval issue might appear on the CSCM agenda and thus endanger the continued presence of the US Sixth Fleet in the Mediterranean. The US administration was also not keen on discussing issues with Libya in such a multinational forum.³⁷ Perhaps, the United States was not fully supportive of the CSCM out of a concern that the United States might lose control of developments in the Middle East. In 1990 neither the United States nor Israel were keen to multilateralize the Middle East problem. There were also apparently divisions in the Arab camp regarding the CSCM, with not all states approving of Egypt's backing.³⁸

The CSCM project was too ambitious, not only in the number of states it intended to cover but also in the range of issues it sought to tackle. The CSCM was proposed at a time when the circumstances were not favorable for such a comprehensive trans-Mediterranean cooperation initiative. The Middle East peace process had not been launched. In order for confidence-building in general to be effective, appropriate political conditions are necessary. The time needs to be "ripe" for a confidence-building process to commence. In 1990 the circumstances in the Mediterranean were not conducive to the launching of open-ended discussions on political and economic reforms between states north and south of the Mediterranean. Cultural differences between peoples needed to be bridged and the sensitivities of Arab governments and even more so Arab publics needed to be addressed. Otherwise, Arab governments and peoples were likely to perceive that the rich and powerful north was attempting to interfere in the internal affairs of the south. In practice, it was not realistic to apply directly, without any modifications, the principles of the CSCE that worked in Europe to the CSCM and a totally different environment in the eastern and southern Mediterranean. Unlike Cold War Europe where tensions were basically caused by an ideological East-West divide, the situation in the Mediterranean was much more complicated. Suspicions between governments and peoples north and south of the Mediterranean and also south-south problems needed to be addressed. According to Stephen C. Calleya, the CSCM was an attempt to establish a trans-Mediterranean international institution that failed because it "attempted to institutionalise regional patterns of amity that [did] not exist."[39]

Serious civil unrest in Algeria and in the Balkans and the UN imposition of sanctions on Libya after the Lockerbie tragedy contributed to the rapid demise of the CSCM project. The CSCM proposal was formally rejected by the EC Council of Ministers at its meeting on May 11, 1991.[40] The CSCE Summit in Helsinki in July 1992 failed to mention the CSCM. The Summit Declaration underlined the need for an "effective information exchange" with the nonparticipating Mediterranean states. It spoke of the importance of widening cooperation and enlarging the "dialogue" with these states to promote social and economic development, enhance stability in the "region," and narrow the gap between states north and south of the Mediterranean.[41] There was no reference to the specific details and goals outlined in the Joint Document at Palma de Mallorca. Spain did attempt to resuscitate the CSCM project by hosting an Inter-Parliamentary Union (IPU) meeting devoted to the CSCM in Malaga in June 1992. Only parliamentary delegations from states that had a Mediterranean coastline were invited to address the meeting. Portugal was thus not present and Algeria was also absent because its legislature had been suspended after the military intervention. Israel and Yugoslavia also failed to send representatives. A status of "associate participant" was offered to the United States, Britain, Russia, and the Palestinians. The IPU Malaga meeting discussed the three baskets of issues proposed at Palma de Mallorca. A document was approved of, but not followed up. There was a proposal to assemble another IPU meeting devoted to the Mediterranean at a later unspecified date.[42] A second gathering would meet in Valletta in November 1995, and a third meeting was scheduled to be held in Tunis in 1999. Parties to "the CSCM process" were also encouraged to debate Mediterranean issues in the two statutory annual conferences of the IPU.[43] However, this was a poor substitute for the original grandiose aims and ambitions outlined in 1990 at Palma de Mallorca.

The French interest in a cooperative arrangement for the western Mediterranean can be traced back to the proposal of President François Mitterrand in January 1983. In a speech in Rabat, Mitterrand called for the establishment of a "Mediterranean Forum"—*Initiative Française en Méditerraneé*—or a Western Mediterranean Conference that would have as its members France, Italy, Spain and Algeria, Morocco and Tunisia. There was also a possibility that Greece and Portugal could join.

However, because of Algeria's lack of enthusiasm the project remained stillborn.⁴⁴

The idea of a cooperative scheme for the western Mediterranean was revived in the late 1980s when Algeria was interested in improving its relations with Europe and more eager to secure a resolution to the problem of the Western Sahara. The Mediterranean Forum concept was discussed in a series of "prediplomatic" meetings at Marseilles (May 1988), Tangiers (May 1989), and Rome (March 1990). Economic, social, and cultural issues were raised.⁴⁵ These meetings laid the preparatory ground for the official inauguration of the Western Mediterranean Forum or "5+4" Group at a gathering of foreign ministers in Rome on December 11, 1990.

In contrast to the CSCM, the Western Mediterranean Forum was more limited in scope, both in the number of states participating and on the nature of issues discussed. The original participants at Rome were France, Italy, Portugal and Spain, and Algeria, Libya, Mauritania, Morocco and Tunisia. The five member states of the AMU were here considered partners in an endeavor to build closer relations between Europe and the Arab world.⁴⁶ The objectives were to work on cooperation in the fields of natural resources management, economic links and financial assistance, migration, and cultural matters. There was no provision for cooperation in the military and defense dimensions of security, although it was declared at Rome that the Western Mediterranean Forum was seeking "to help transform the Mediterranean into an area of peace and cooperation."⁴⁷ At Rome it was decided to create a Mediterranean Data Bank. There was also talk of possible joint work in the fields of anti-drug-trafficking and the delimiting of national territorial waters and airspace. The intention was to hold regular ministerial meetings and set up working groups to tackle various economic and social problems.

The second ministerial meeting of the Western Mediterranean Forum assembled in Algiers in October 1991. By this time Malta had been formally included and the "5+4" Group had become the "5+5" Group. At Algiers the participants spoke of the prospects of creating a Mediterranean Bank for Industry and Commerce, and a Mediterranean Financial Bank. They also talked of the importance of cooperating in the areas of science and technology. It was agreed to set up a working group to examine migration issues, and a meeting of employment and labor ministers was proposed for some time in 1992. The parties also decided to hold a summit meeting of their heads of state and government in Tunis in early 1992.⁴⁸

Given Italy's interest in the Mediterranean in its broadest sense, it was not surprising that at the Algiers meeting Italian officials had pressed for Greece and Egypt to be included as observers in the Western Mediterranean Forum.[49] The Tunis meeting was never held and the "5+5" Group collapsed. The crisis in Algeria and the Lockerbie disaster made it virtually impossible for the Europeans to consider the AMU a reliable negotiating partner. Events in the Middle East and a corresponding shift in the EU's policy toward the Mediterranean also made the Western Mediterranean Forum appear outmoded if not irrelevant.

For a brief period the Western Mediterranean Forum was a rival to the CSCM project and may have contributed toward the early demise of the CSCM initiative. In the early 1990s, France, Italy, and Spain clearly had divergent aims and interests in the Mediterranean. The Western Mediterranean Forum had, in effect, attempted to focus on issues relating to the second basket of the CSCM. But even this less ambitious trans-Mediterranean cooperation initiative fell victim to changing circumstances in the area. Nevertheless, in later more favorable conditions, other initiatives and trans-Mediterranean dialogues would commence that would learn from the experiences of the aborted CSCM and Western Mediterranean Forum. The latter project had aimed to focus cooperation at the intergovernmental level. The CSCM concept had also aimed to establish direct contacts with peoples through its human dimension. It appears that the CSCM idea was ahead of its time. By the mid-1990s new initiatives, such as NATO's Mediterranean dialogue, were attempting to take advantage of the favorable turn of events in the Middle East and strike up links—albeit tentatively—not just with Arab governments but also with Arab publics.

THE SIGNIFICANCE OF NATO'S STRATEGIC CONCEPT

Speaking at the Munich Conference on Security Policy on February 5, 1995, the US secretary of state for defense, William Perry, declared: "We must all come to grips with the threats to our interests posed by the growth of instability and extremism in north Africa and elsewhere. This is not just a southern European problem."[50]

A few weeks later, on February 27, in a speech at AFSOUTH headquarters in Naples, the US assistant secretary of defense for European and NATO Affairs, Joseph Kruzel, noted: "Today the real threat to European security comes not from the northern region, where much of the Alliance's attention is now focused, but in the south, where existing conflicts and potential for catastrophe are pervasive."[51]

On the face of it, such stark references in early 1995 to a threat from the south appear surprising. This was more like the language of 1990 and 1991 at the time of the Iraqi invasion of Kuwait. For example, an annual public-opinion poll conducted in France in May–June 1991 revealed that 59 percent thought the principal threat to France came from the south as opposed to only 8 percent who thought it stemmed from the east. In this poll the most feared countries included Libya and Algeria, immediately behind Iran and Iraq.[52] By November 1991, however, talk in NATO circles of a threat from the south had almost become passé. NATO heads of state and government meeting in Rome issued the Alliance's new Strategic Concept for the post–Cold War era. In addition to its references to the broadened concept of security, the Strategic Concept referred more to "risks" and "challenges" than threats from territory traditionally regarded as out-of-area.

According to the Alliance's new Strategic Concept, NATO was now confronted with "security challenges and risks" in place of the "predominant threat" that the Soviet Union had posed (paragraphs 8–9). These risks were more difficult to predict and assess than the past Soviet threat, as they were "multi-faceted in nature and multi-directional" (paragraph 9). The risks were of a "lesser magnitude" than the previous Soviet threat, but different crises were more likely to arise "which could develop quickly and would require a rapid response" (paragraph 14). Risks to the security of NATO member states were less likely to result from "calculated aggression" but "rather from the adverse consequences of instabilities that may arise from the serious economic, social and political difficulties" (paragraph 10).[53]

The Strategic Concept referred extensively to challenges and risks that may stem from central and eastern Europe and the Soviet Union, but the Mediterranean, the Middle East, and the Gulf were also mentioned. Paragraph 12 noted that the Gulf War of 1991 demonstrated that the "stability and peace of the countries on the southern periphery of Europe are important for the security of the Alliance." Attention was drawn to the dangers of the proliferation of WMD and ballistic missiles in the southeastern Mediterranean and in the Middle East. The Atlantic Alliance sought to maintain "peaceful and non-adversarial relations with the countries of the southern Mediterranean and in the Middle East." Paragraph 13 spoke of "risks of a wider nature" that could entail terrorism and sabotage, as well as the disruption of the flow of vital resources. Certainly, the Mediterranean could be included in this context. The Strategic Concept did link together risks and threats by noting that risks could become threats to the security of individual members, or the Atlantic Alliance as a whole, if left unchecked. Preventive diplomacy—crisis anticipation and prevention—was thus underlined, in order to tackle potential and actual risk situations before they could become threats (paragraph 32).[54] Only days earlier Secretary-General Wörner in a speech at Madrid had advised that risks should not be "contained" but rather "tackled at source."[55]

According to the standard dictionary definition, a "threat" may involve the act of declaring an intention to inflict pain. Wörner had wisely noted that a threat involves "a combination of capability and intention."[56] Intention alone is thus of little consequence. Without the capability to carry out a threat, there is no threat as such. At best, in such circumstances the threat is only potential and not real. As noted in the Strategic Concept, unlike threats, risks are not premeditated. Risks are rather the manifestations of instability. However, if left unchecked, risks may become threats.

Perceptions must also be considered. In the Cold War all NATO allies had perceived that there was a Soviet threat. The threat was massive, observable, and real. In contrast, risks are more difficult to quantify. All NATO member states may be able to perceive manifestations of instability, but it is less easy for NATO allies to reach the common conclusion that these risks may become or actually have become threats. The distinction between risks, potential threats, and actual threats is not clear-cut. Capability is easier to ascertain than intention. And, of course, intentions can always change.

As noted previously, governments and publics in the southern Mediterranean do not look favorably on being regarded as a threat to the north. Indeed, in the south the talk is at times about the threat posed by the north to the south. References in NATO circles to WMD, terrorism, and Islamic radicalism/fundamentalism all sound "threatening." But, as discussed earlier, there are manifestations of instability in the southern and eastern Mediterranean caused by political, economic, social, and environmental problems that need immediate addressing. Cooperation between governments and peoples north and south of the Mediterranean on these issues would help build confidence in the region. This cooperation could only come about and succeed if governments and peoples around the Mediterranean not look upon each other as potential or actual threats. Certain prerequisites are thus needed before trans-Mediterranean cooperation initiatives can be developed. Parties must have "incentives" to embark on a negotiating process. The costs of possible war or heightened tensions should be acknowledged. Most important, a minimum degree of trust between the parties concerned is essential.

The Strategic Concept is also of significance because of its several references to the value of dialogue. These references were actually addressed to governments in central and eastern Europe and in the Soviet Union, but they should also be borne in mind for the later NATO-Mediterranean dialogue. The broad approach to security, and the problem of risks outlined in the Strategic Concept, led NATO policy-makers to emphasize the crucial role of dialogue. In the text it was noted that the Atlantic Alliance's active pursuit of dialogue and cooperation, underpinned by a commitment to collective defence,

> seeks to reduce the risks of conflict arising out of misunderstanding or design; to build increased mutual understanding and confidence among all European states; to help manage crises affecting the security of the Allies; and to expand the opportunities for a genuine partnership among all European countries in dealing with common security problems. (paragraph 26)[57]

In 1967 the Harmel Report had underlined the importance of dialogue with Warsaw Pact states at the height of the Cold War. The Strategic Concept called for enhancing this dialogue in the post–Cold War era and spoke of the need to build a partnership between states across Europe. This would soon result in the formation of the NACC, a new multilateral forum for dialogue and cooperation between European states. The Strategic Concept also positively made note of the "growth of freedom and democracy" in eastern Europe, which multiplied the opportunities for dialogue (paragraph 29).[58] The implication, here, was that a dialogue would be more difficult to conduct with states that are not democratic.

The Strategic Concept also linked dialogue and cooperation with crisis management and conflict prevention. Fully developed dialogue and cooperation within Europe would "help to defuse crises and to prevent conflict" (paragraph 34). There were several references to crisis management and conflict prevention (paragraphs 32–34), and the text also stated that NATO could be employed "as an instrument of crisis management" (paragraph 47).[59] In the following months there was much talk in NATO circles of the need for new "core functions" for the Atlantic Alliance in addition to the traditional emphasis on collective defence. "Crisis management" and the value in "projecting stability" to former adversaries were listed as examples of new core functions.[60] Projecting

The Immediate Background to NATO's Mediterranean Initiative 143

stability was, in effect, a form of conflict prevention. The NACC became an important vehicle for projecting stability to former enemies. The involvement of NATO in the Balkans was an illustration of the Atlantic Alliance's newfound interest in crisis management in territory traditionally regarded as out-of-area.

The Final Communiqué of the NAC meeting in Athens in June 1993 declared that: "Conflict prevention, crisis management and peacekeeping will be crucial to ensuring stability and security in the Euro-Atlantic area in the years ahead."[61] There was no specific mention of the Mediterranean in this direct reference to the applicability of new core functions in territory that was once regarded as out-of-area. However, as will be discussed later, elsewhere in this communiqué there was a reference to the connection between the security of Europe and the Mediterranean and the importance of dialogue and cooperation. One may contend, therefore, that by implication, the Atlantic Alliance's interest in crisis management and conflict prevention could be extended to apply also to the Mediterranean.

In the months following the November 1991 Rome Summit, the attention of NATO focused on relations with central and eastern Europe and what became the states of the former Soviet Union. The NACC was established at the end of 1991. The NATO Summit of January 1994 announced the PfP program. The Atlantic Alliance began to wrestle with the idea of enlarging NATO to its east without alarming Russia. Special arrangements would be worked out between NATO and Russia, and NATO and Ukraine. In line with the notion of an ESDI, NATO began to coordinate its relations with WEU. The concept of Combined Joint Task Forces (CJTFs), formally introduced at the NATO Summit of January 1994, would enable NATO assets to be used in a possible future WEU-led military operation and would make it easier for non-NATO countries to participate in peacekeeping deployments with NATO units. Discussions were also under way concerning how to reorganize NATO command structures. In vain, France attempted to persuade the United States to relinquish its control of AFSOUTH. There was also a debate about whether AFSOUTH should be elevated in status and become a major command instead of its current subordinate command status. It was argued that the establishment of an Allied Command Mediterranean as a new, and third major command, would be a recognition of the increased importance of security issues to NATO's south.[62]

In spite of references to the Mediterranean with regard to the reorganization of NATO command structures, there was actually little interest in Atlantic Alliance circles on problems in the southern and eastern Mediterranean in the months immediately after the 1991 Rome Summit. The Club Med grouping of NATO allies was unable to divert the attention of other member states from what were perceived to be more important priorities. However, notwithstanding the failings of the CSCM and the Western Mediterranean Forum, there were other ongoing dialogues and initiatives that the Club Med states could concentrate on after 1991. And by mid-1993, NATO itself began to reassume a heightened interest in events in the Mediterranean, as the risks, challenges, and potential, if not real, threats in the area did not disappear.

THE BEGINNING OF A REAL DIALOGUE WITH THE MEDITERRANEAN

The failed attempts at trans-Mediterranean cooperative initiatives in the late 1980s and early 1990s had predated the important breakthrough in the Middle East peace talks. From 1992 onward the United States increasingly had to take into account the enhanced interest of western Europeans and the international community in general in what was now the Middle East peace process. There were better prospects for including states from the eastern Mediterranean as well as from the southern Mediterranean in a north-south dialogue. Potentially successful trans-Mediterranean cooperation initiatives no longer necessarily needed to gravitate around ties between the AMU and a select number of northern Mediterranean states.

The Mediterranean Forum, which is also known as The Forum for Dialogue and Cooperation in the Mediterranean, was officially launched in Alexandria in July 1994. France, Italy, and also Egypt promoted this new trans-Mediterranean cooperation initiative. Egyptian officials had been frustrated in their efforts to join the AMU and hence, much to their disappointment, had been excluded from the Western Mediterranean Forum or "5+5" Group. Egypt was eager to establish closer ties with the EU. The authorities in Cairo perceived that they were losing a grip on regional leadership with positive developments in the Middle East. They thus appeared keen to add a new Mediterranean dimension to Egyptian foreign policy and thereby hoped to gain better access to EU economic resources and support.[63]

After failing to secure admission to the Western Mediterranean Forum, in 1991, in a speech to the Council of Europe, President Mubarak first floated the idea of a broader, and not exclusively western, Mediterranean grouping. Egyptian officials were determined to include traditional security interests and also terrorism within the remit of this broader grouping. By November 1993 Egypt had submitted a draft plan of action in this regard to various interested countries.

Seeking to resuscitate, in some form, its own expiring Western Mediterranean Forum, the French government was willing to cosponsor the Egyptian proposal. The Italian government also looked favorably on the Egyptian initiative and requested the Institute of International Affairs in Rome to prepare a report about what forms of Mediterranean cooperation could be achieved in a new and enlarged Mediterranean Forum. The Mediterranean 2000 Report was thus adopted as the basis for discussion at the inaugural meeting of the Mediterranean Forum in Alexandria.

The foreign ministers of France, Greece, Italy, Portugal, Spain, Turkey, Algeria, Egypt, Morocco, and Tunisia attended the Alexandria meeting on July 4, 1994. Malta would also later be admitted to the new Mediterranean Forum. Only Mediterranean littoral states (with the exception of Portugal) could be members, but Israel, Libya, and Syria, for example, were not participants. The Alexandria Communiqué spoke of governmental and nongovernmental cooperation in certain spheres. It was decided to set up three working groups—on political dialogue, on a dialogue between cultures and civilizations, and on economic and social activities. A timetable of ministerial and technical meetings was agreed upon. There was talk of the need for practical action programs in such fields as culture, education, emigration, employment, energy, industrial development, and the environment. One aim was to coordinate the work of universities, research institutions, associations, trade unions, and companies based around the Mediterranean. The ultimate objective was genuine "Mediterranean political cooperation." Traditional security concerns would be addressed in periodic reports on "the status of cooperation and conflicts" in the region, prepared by a network of institutes.[64]

Although not an exclusively western Mediterranean body, Syria and Israel were intentionally not invited to join the Mediterranean Forum until further progress was made in the Middle East peace process. Libya remained a pariah state. The cosponsors of the Mediterranean Forum did not want the success of their new grouping to become tied in with and dependent on events in the Middle East. But this meant that the Mediterranean Forum was not able to tackle security issues for the Mediterranean as a whole. Thus, the title, Mediterranean Forum, was in effect a misnomer.

The working groups on political, cultural, and economic issues assembled for the first time in October 1994 in Madrid, Rome, and Cairo respectively. In December 1994 senior officials met in Algarve to consider the initial activities and programs of these groups and prepare for a ministerial meeting in Sainte Maxime in southern France to be held in April 1995. But, already, the Mediterranean Forum was in danger of becoming redundant, as the EU by this time was seriously considering the launching of a new Euro-Mediterranean Partnership.

In the wake of earlier aborted trans-Mediterranean cooperation initiatives, established institutions and organizations also launched or further developed dialogues between states north and south of the Mediterranean. In line with a decision taken at its Helsinki Summit of July 1992, the CSCE arranged a seminar devoted to the Mediterranean in Valletta in May 1993. Egypt, Israel, Lebanon, Libya, Morocco, and Tunisia attended and contributions were made by Egyptian, Israeli, Moroccan, and Tunisian delegates. In his opening statement the deputy prime minister and foreign minister of Malta, Professor Guido de Marco, argued in favor of setting up a Council of the Mediterranean modeled on the Council of Europe. Environmental issues and the migration question were discussed. Officials spoke of the need for a "wider and more regular political dialogue." The nonparticipating Mediterranean states requested "more regular consultations" and expressed an interest "in being associated with the CSCE in a more permanent and structured relationship."[65]

In May 1993 Morocco called to be admitted as an observer to the CSCE. The following month Egypt appealed for the establishment of a more structured and institutionalized relationship between the CSCE and Egypt.[66] In the face of this pressure, in September 1993 the Committee of Senior Officials of the CSCE made an announcement. The chairman-in-office of the CSCE was requested in consultation with the CSCE troika—consisting of the preceding, current, and succeeding chairmen-in-office—and the nonparticipating Mediterranean states to make proposals for an "increased dialogue" between the CSCE and those nonparticipating Mediterranean states "which share its principles and objectives."[67] In March 1994, at a meeting of the Committee of Senior Officials, Algeria, Egypt, Israel, Morocco, and Tunisia were listed as five nonparticipating Mediterranean states that shared the principles and objectives of the CSCE. In future, they would be "regularly invited" to meetings of the CSCE Council of Ministers, CSCE Review Conferences, meetings of the CSCE troika, and other seminars and ad hoc gatherings. Participation in these meetings would be reviewed, though, bearing in mind the extent to which the five listed states "continue to share CSCE principles and objectives."[68] These principles included respect for human rights and fundamental freedoms, the self-determination of peoples, and the need for free, open, and competitive elections. Perhaps not surprisingly, therefore, Lebanon, Libya, and Syria were not listed as states entitled to attend future CSCE meetings. The inclusion of Algeria, however, was certainly controversial bearing in mind the imposition of military rule in that country in 1992. In practice, the implied threat that nonparticipating Mediterranean states could be refused admission to CSCE gatherings for bad behavior has turned out to be a hollow one.

In December 1994 at the landmark Budapest Summit of the CSCE—which was renamed the OSCE—decisions were taken to enhance the dialogue with the Mediterranean. This was evidently due to Egyptian pressure.[69] The OSCE participating states were to establish an "informal open-ended contact group at the level of experts" within the framework of the OSCE Permanent Council in Vienna. This group would meet periodically with representatives of the five nonparticipating Mediterranean states to conduct a dialogue and facilitate the exchange of information. The OSCE chairman-in-office, with other members of the troika and the secretary-general, would also conduct "high level consultations" with these nonparticipating Mediterranean states. The chairman-in-office could also invite representatives of these states to attend meetings of the Senior Council when Mediterranean issues were on the agenda, and also to meetings of the Forum for Security Cooperation and the Permanent Council when Mediterranean issues were to be discussed. Building on the tradition of past CSCE seminars, an OSCE seminar would be organized to discuss the CSCE experience in the field of CBMs.[70]

Italy and Malta had proposed that OSCE "associate member" status should be granted to those nonparticipating Mediterranean states willing to abide by OSCE principles.[71] But, the rather unflattering title of nonparticipating Mediterranean states remained.

The CSCE/OSCE-Mediterranean dialogue had evolved from one that had consisted of a very modest and unstructured occasional series of talks with various "Mediterranean" states to one that had become much more organized and institutionalized with a select list of countries from the western and eastern Mediterranean. The OSCE dialogue was in no way comparable, though, to the scale and ambition of the previously proposed CSCM. The OSCE encouraged the five nonparticipating Mediterranean states to attend briefings as a group. But, in the more formal negotiating sessions, OSCE officials would meet separately with representatives from each of the nonparticipating Mediterranean states. Morocco and Egypt in the months prior to the Budapest Summit had been consistently lobbying for the CSCE to focus more attention on the Mediterranean. By the end of 1994, though, it was not clear how effective the enhanced OSCE-Mediterranean dialogue would be. Some of the problems that had been raised with regard to the CSCM again needed to be considered. In spite of the references to nonparticipating Mediterranean states ostensibly sharing the principles and objectives of the CSCE/OSCE, in practice just how relevant for the southern and eastern Mediterranean was the CSCE experience in Europe? For example, the Budapest Summit publicized a new Code of Conduct on Politico-Military Aspects of Security that emphasized the importance of civilian control over military, paramilitary, and other security forces. In reality, as noted previously, this Code of Conduct could scarcely be applied in most of the nonparticipating Mediterranean states.

The idea of a Council of the Mediterranean raised at the CSCE Seminar in Valletta in May 1993 had stressed the value of the rule of law and the respect for the dignity of individuals. Maltese Deputy Prime Minister de Marco had suggested that the Council, with a Committee of Ministers and a general assembly with consultative powers that might become a Parliamentary Assembly of the Mediterranean, could become involved in political, economic, social, environmental, and cultural issues. However, apparently, the Maghreb states only offered lukewarm support.[72] In part, this may have been because none of the major European states were eager to back the Maltese initiative.

Little has emerged from the Maltese proposals. The Parliamentary Assembly of the Council of Europe has considered initiating contacts with those nonmember Mediterranean countries that apply the principles of parliamentary democracy, respect human rights, and the rule of law.[73]

Unless there is substantial political reform in these states, the Council of Europe will likely not pay too much attention to the Mediterranean. Thus, the Council of Europe will probably have little impact on developments in the area.

WEU had also commenced a dialogue with the Mediterranean before NATO launched its Mediterranean Initiative. In the shadow of NATO, throughout most of the Cold War, WEU had been a moribund organization. In 1984 WEU was revitalized, as the United States was eager for the Europeans to shoulder more of the defense burden, and the Europeans in turn were less willing to tow the aggressive anti-Soviet line adopted by the administration led by Ronald Reagan. In October 1987 at

The Hague, WEU foreign ministers and defense ministers issued a declaration known as the "Platform on European Security Interests." Ministers decided to "concert" their "policies on crises outside Europe" insofar as they might affect the security interests of WEU member states.[74] Therefore, for example, immediately after the Iraqi invasion of Kuwait in 1990, WEU member states agreed to coordinate naval operations in the Gulf. In the Bosnian crisis, ships of NATO and WEU in the Adriatic enforced UN sanctions. WEU also participated in enforcing sanctions on the Danube. In 1994 the EU and WEU established a joint administration of the city of Mostar in order to monitor public order and help create a Bosnian-Croat local police.

By the early 1990s there was much discussion about the relationship between WEU and the EU, and about what role WEU could play together with NATO in an ESDI. The December 1991 Maastricht Treaty of European Union noted that the EU might request WEU, "which is an integral part of the development of the Union, to elaborate and implement decisions and actions of the Union which have defence implications."[75] In what amounted to a compromise, it was also noted that "WEU will be developed as the defence component of the EU and as a means to strengthen the European pillar of the Atlantic Alliance."[76]

WEU member states were also assuming an increased interest in possibly participating together in operations in areas beyond the territory of member states—in territory that was out-of-area for WEU. The June 1992 WEU Petersberg Declaration declared that military units of WEU member states acting under the authority of WEU could be employed for humanitarian and rescue tasks, peacekeeping roles, and tasks of combat forces in crisis management including peace-making.[77] Significantly, no geographic limitation was placed on these potential operations. In practice, though, unlike NATO, WEU lacked the logistical and airlift capability to mount a substantial and prolonged military operation out-of-area. However, in theory, a WEU-led CJTF making use of NATO (that is, the United States) resources and assets could act out-of-area. By late 1992 Italy was lobbying for the creation of a multilateral ground force and air-maritime unit that could be deployed in the Mediterranean.[78] Eventually, this would lead to the formation of EUROFOR and EUROMARFOR as forces answerable to WEU. In 1992, 1993, and 1994 French, Italian, and Spanish forces held joint exercises in the Mediterranean.

The admission of Portugal and Spain to WEU as full members in 1990, and then of Greece in 1995, and the appointment of Portuguese ambassador Jose Cutiliero as WEU secretary-general in 1994, gave

WEU a more Mediterranean focus. Since July 1987 the Permanent Council of WEU had been overseeing the activities of a WEU subgroup of experts on the Mediterranean that had been studying a range of issues related to security. In 1993 the subgroup was upgraded to become a Council working group. The WEU Petersberg Summit of June 1992 heralded the commencement of a dialogue between WEU and the Mediterranean. The presidency and secretariat of WEU were to begin holding twice-yearly meetings with representatives of the embassies (in London, and then in Brussels when WEU moved its headquarters) of certain "non-WEU Mediterranean countries." These first contacts were diplomatic in nature. The representatives of non-WEU Mediterranean countries were received individually by leading WEU officials. The dialogue between the Mediterranean and WEU was meant to be a "gradual and phased dialogue with the Maghreb countries, taking into account the political developments both in the countries and in the region."[79] Contacts would begin with Algeria, Morocco, and Tunisia, three members of the AMU that had been developing links with the EU. Mauritania would be included later in the dialogue, but Libya was not invited. Egypt also entered the dialogue in May 1994 to be followed by Israel and also Cyprus and Malta in 1995.[80]

In addition to this diplomatic dialogue, by autumn 1994 the WEU Mediterranean Group had also started to hold separate sessions with government experts from Algeria, Egypt, Mauritania, Morocco, and Tunisia. These were supposed to be organized twice yearly. The WEU Institute for Security Studies based in Paris was encouraged to work closely with the WEU Mediterranean Group. In December 1994 the Institute was tasked by the Permanent Council of WEU to analyze the security and defense policies of the Maghreb countries and Egypt in liaison with the security institutes in those countries.

The WEU-Mediterranean dialogue, both at the level of experts and at the diplomatic level, was based on seven principles. These were: transparency in military activities and military doctrine; the emulation of CSCE/OSCE mechanisms for consultation, confidence-building, and fact-finding; conflict prevention; sufficiency in conventional armed forces; nonproliferation; the peaceful settlement of disputes; and common security perceptions concerning the region.[81] The focus was thus on the military aspects of security.

However, bearing in mind the concerns of Arab ruling elites with regard to issues of societal security, the non-WEU Mediterranean countries also wanted to include social, economic, and even cultural issues in the

WEU-Mediterranean dialogue. WEU officials argued that the OSCE and the EU should address such issues.[82]

It is quite possible that NATO officials would have heeded the complaints of the non-WEU Mediterranean countries and, in line with NATO's Strategic Concept, employed a broader concept of security when embarking later on the Atlantic Alliance's Mediterranean Initiative. The WEU-Mediterranean dialogue would also run into difficulties as doubts resurfaced concerning the future importance and relevance of WEU, particularly with regard to the EU and the problems in shaping a Common Foreign and Security Policy for Europe.

Although the Euro-Maghreb dialogue was largely a failure, the EU was involved in other initiatives with non-EU Mediterranean countries that would eventually culminate in the convening of the Barcelona Conference in November 1995. Since the late 1960s the western Europeans—through the European Economic Community (EEC) and then the EC—had concluded a series of bilateral agreements with Mediterranean states providing technical and financial assistance and favorable terms of trade. Agreements were made with Algeria, Egypt, Israel, Jordan, Lebanon, Morocco, Syria, and Tunisia. Financial aid remained relatively modest even though five yearly financial protocols were introduced in 1976. The negative impact on North Africa and the eastern Mediterranean of the admission of Portugal and Spain to the then EC in 1986, and of the Single European Act, which entered into operation on January 1, 1993, have been previously noted.

In December 1990 the EC announced its Renovated Mediterranean Program (RMP) for the period 1992–96. This envisioned financial assistance of over ecu 4 billion in loans and gifts. This was triple the sum offered for the period 1987–91. Nevertheless, the aid package to the Mediterranean was worth only ecu 2.4 per capita per year compared to ecu 6.8 per capita per year for central and eastern Europe.[83] The RMP called for economic liberalization and structural adjustment reforms in North Africa and the Middle East. Significantly, in 1992 the EC was allocated the chair of REDWG, one of the multilateral working groups established as part of the Middle East peace process. But by 1992 the RMP was coming under increasing criticism from non-EU countries on the grounds that it was not doing enough. Thus, the European Council in Lisbon in June 1992 called for an expansion of relations with countries to the south of the EU. In particular, the European Council urged that a Euro-Maghreb "partnership" based on "codevelopment" should be established that would encompass trade liberalization, political dialogue, and cooperation in the social and cultural fields and also on questions of security.[84] There were also calls at Lisbon for a Euro-Maghreb free trade area.

Continued problems with Libya and the worsening crisis in Algeria hindered the cultivation of an exclusively EU-Maghreb axis.[85] Still, by April 1993 the EU's European Commission proposed that new agreements be negotiated with Algeria, Morocco, and Tunisia on the basis of "partnership" and integrated cooperation. Provisions for an industrial free-trade regime were to be included in these agreements. As a result, new association agreements would be concluded in 1995 between the EU and Morocco and Tunisia in the months before the Barcelona Conference.[86]

It is also worthy of note that in 1992 the EU had set in motion the promotion of cooperation between civilian bodies in order to build up a network of links between institutions and organizations on both sides of the Mediterranean. Consequently, connections were established between municipal authorities (MED-URBS), universities (MED-CAMPUS), journalists (MED-MEDIA), and businessmen (MED-INVEST).

The EU heads of state and government, assembled at Corfu in June 1994 for a meeting of the European Council, mandated the EU Council of Ministers and the European Commission to consider how to strengthen further EU-Mediterranean links. Pressure was now being applied not only by the southern Mediterranean countries but also by the states of southern Europe who were increasingly concerned at what they perceived to be a disproportionate amount of aid being channeled to central and eastern Europe.[87]

Hence, on October 19, 1994, the European Commission proposed that a Euro-Mediterranean partnership—rather than a more restricted Euro-Maghreb partnership—be encouraged. The intention was to establish a zone of stability and a free-trade area by the year 2010. The Commission recommended that ecu 5.5 billion should be offered to support the EU's Mediterranean policy. The Commission also proposed that the European Investment Bank offer the same amount.[88]

The meeting of the European Council in Essen on December 9–10, 1994, took into account the Commission's suggestions and decided to hold a conference in Barcelona to discuss political, economic, financial, human, societal, and security (in the traditional sense) issues related to the Mediterranean. The Essen Summit noted: "The Mediterranean represents a priority area of strategic importance for the EU." It was proposed that invitations be extended to Algeria, Cyprus, Egypt, Israel, Jordan,

Lebanon, Malta, Morocco, Syria, Tunisia, Turkey, and the occupied territories.[89] Libya was omitted. Mauritania was also excluded, as it was considered a part of sub-Saharan Africa. It seems that as well as pressure from states of the southern Mediterranean and southern Europe, positive developments in the Middle East peace process had given added momentum to the EU's Mediterranean policy. NATO officials were most probably influenced by the EU's heightened interest in the Mediterranean.

The work of the NAA—the body of parliamentarians from NATO member states that has close ties with the Atlantic Alliance but is not officially affiliated to it—also had a major influence on NATO officials and their interest in the Mediterranean. Following up on earlier work on the out-of-area debate, to discuss security issues pertaining to the Mediterranean and the Middle East in 1990, the Civilian Committee of the NAA formed a Sub-Committee on the Mediterranean Basin, and the Political Committee of the NAA formed a Sub-Committee on the Southern Region. Delegations from these subcommittees visited Egypt and Israel in 1992, and Morocco in 1993. Between 1992 and 1994 a series of Rose-Roth Seminars were organized by the NAA, which were devoted to the Mediterranean and were attended by representatives from non-NATO states of the southern and eastern Mediterranean.

At its Oslo meeting in May 1994 the Standing Committee of the NAA decided to initiate a dialogue with parliamentarians from the southern Mediterranean with the aim of enhancing regional security. This dialogue would assume the form of an annual Mediterranean seminar. The first of these seminars would actually only convene in March 1995 in Paris, one month after the announcement of the launching of NATO's Mediterranean Initiative. Parliamentarians from Cyprus, Egypt, Israel, Jordan, Malta, Morocco, and Tunisia participated together with a delegation of Palestinians. Deputies from Lebanon and Syria declined an invitation to attend. Parliamentarians from Algeria and Mauritania were not invited.[90] In order to handle problems of instability in the Mediterranean, the seminar stressed that economic rather than military measures of support were required. At Oslo in May 1994 the NAA granted Morocco observer status. Israel became an observer in November 1994 and Egypt in May 1995. Tunisia participated on an ad hoc basis. Observer status was to be accorded to those states that were perceived to be "emerging as democracies."[91]

Reports of the NAA published in 1992 pressed for the beginning of a dialogue between states north and south of the Mediterranean. Bruce George advised that a dialogue with the southern Mediterranean should parallel as appropriate the activities of the NACC in central and eastern Europe, with a focus on issues relating to defense, security (in its traditional sense), and political, economic, and ecological issues. George noted that this dialogue could take the form of a CSCM or some other subregional, trans-Mediterranean coalition.[92] No specific reference was made to NATO acting as a vehicle for this dialogue. On the other hand, in his 1992 report Miguel Herrero stated that the "increased political dimension of the Alliance enables the establishment of a dialogue with states and structures in the South as well as in the East." Herrero emphasized that there was a need to build a dialogue with the south that should concentrate on security and arms control.[93]

A special NAA study published in September 1993 suggested that NATO should develop a version of the NACC for the south to promote a security dialogue. This would be part of a policy to make NATO more active, informed, and forward looking.[94]

One month later the NAA was urging NATO to launch an "active" outreach program directed toward the southern Mediterranean Basin that could assist in facilitating dialogue and cooperation with a view to enhancing regional security.[95]

NATO DIRECTS ITS ATTENTION TOWARD THE MEDITERRANEAN

Given the developments outlined in this chapter it is not surprising that NATO itself should finally commence its own dialogue with the Mediterranean in February 1995. It would have looked odd for NATO to ignore the Mediterranean given the interest, for example, of the NAA, the EU, the OSCE, and WEU in the south. But what should constitute a NATO-Mediterranean dialogue? A consensus would have to be reached between NATO member states on such questions as which countries should be invited, what issues should be discussed, and at what level the dialogue should be conducted.

Apparently, Canada and the north European members of the Atlantic Alliance only threw their weight behind the initiative when they were assured that the exercise would be cost free, would remain at the diplomatic level for the foreseeable future, and would not divert NATO's attention away from central and eastern Europe.[96] Evidently, not all members of the Atlantic Alliance were overly apprehensive about a threat from the south. Perhaps there was rather more concern that the southern members of NATO should not feel marginalized. Certainly,

Spain, strongly backed by Italy and Portugal, was in favor of embarking on the Mediterranean Initiative. NATO may not have fragmented on this issue, but it seems that northern members of the Atlantic Alliance realized the importance of supporting its southern members on what at the time looked like a relatively low key diplomatic exercise. One prominent American analyst of NATO noted in early 1995, though, that the United States had become more interested in cultivating a dialogue with the Mediterranean to enable NATO to adapt to new security challenges. In the words of this analyst, the dialogue with the Mediterranean was a "significant step."[97]

Until the communiqué released after the meeting of the NAC in Athens in June 1993, there had been no reference to the Mediterranean in official NATO texts since the publication of the Alliance's Strategic Concept. The Athens communiqué noted for the first time that: "Security in Europe is greatly affected by security in the Mediterranean." This formula would be repeated almost ad nauseam in future NATO gatherings. At Athens it was also stated with regard to the Mediterranean that NATO would encourage "all efforts for dialogue and cooperation which aim at strengthening stability in the region. The example of our improved understanding and cooperative partnership with the countries of central and eastern Europe would serve to inspire such efforts."[98] This was the first explicit mention in an official NATO text of the importance of dialogue with the Mediterranean, although the need for a NATO-Mediterranean dialogue was not specified. The Athens communiqué may have been referring to ongoing north-south Mediterranean initiatives and dialogues. It was not made clear how the NACC could serve as a model for future cooperation with the Mediterranean, although the communiqué here was, in part, echoing previously mentioned NAA reports that had referred to the relevance of the NACC for NATO's relations with the south.

Surprisingly, the NAC communiqué released in December 1993 in Brussels contained no reference to the Mediterranean even though the Brussels NATO Summit was only a few weeks away. Due to fierce lobbying from Italy and Spain, the Brussels Summit did refer to the Mediterranean, albeit briefly. In addition to noting how security in Europe was affected by security in the Mediterranean, the Summit also declared that breakthrough in the Middle East peace process had opened the way "to consider measures to promote dialogue, understanding and confidence-building between the countries of the region." The Council in Permanent Session was directed to continue to review the overall situation while NATO encouraged "all efforts conducive to strengthening regional stability."⁹⁹ Thus, there was only a passing reference to the Mediterranean as a whole in a paragraph that was largely devoted to considering developments in the Middle East. Again, NATO did not commit itself to embarking on any dialogue with states of the Middle East and the Mediterranean. In his analysis of the Brussels Summit, George noted that there were few references to the Mediterranean because of continued differences NATO as an organization had with the issue of the Mediterranean. French officials were not sure if NATO had any business in Africa while the United States was still concerned that the Atlantic Alliance could create difficulties for the Middle East peace process and could perhaps endanger the continued presence of the US Sixth Fleet in the Mediterranean.¹⁰⁰

It has been suggested that in the course of 1994, NATO officials became increasingly alarmed about the situation in North Africa and the Middle East, with the continuing civil unrest in Algeria. Arab states—especially Egypt—were reluctant to renew the NPT, which could result in the proliferation of other types of WMD in the area.¹⁰¹ NATO's increased interest in the danger of weapons proliferation may have made more NATO member states more conscious of a possible threat from the south. Certainly, at the meeting of the NAC in Istanbul in June 1994, it was stated that NATO was carefully following political developments around the Mediterranean. The communiqué added: "We are concerned by the risks to stability in this area." The Council in Permanent Session was instructed to continue to review the overall situation and "to examine possible proposals by its members with a view to the strengthening of regional stability" in the Mediterranean.¹⁰²

The Brussels Summit had not referred to "risks" although "risks" and "threats" had been extensively discussed at the Rome Summit only just over two years ago when the Alliance's Strategic Concept was publicized. The concern for proliferation was probably a key reason for the reference in Istanbul once again to "risks." The communiqué in Istanbul was also a considerable advance on the much more ambiguous statement of the Brussels Summit in that reference was made to NATO member states possibly making proposals with regard to the Atlantic Alliance's policy to the Mediterranean. This would have encouraged the "Club Med" members of NATO to recommend the launching of a form of dialogue with the Mediterranean.

Mediterranean security was one of the four topics addressed at the informal meeting of NATO defense ministers in Seville on 29–30 September 1994. Apparently, this was why at Seville a French defense minister attended an official NATO gathering for the first time since 1966.[103] French officials had by now come to the conclusion that NATO should be more involved in the Mediterranean. At Seville the French defense minister proposed the formation of a "Partnership with the South." The US administration, likewise, had assumed a greater interest in NATO's involvement in the Mediterranean. The US Secretary of Defense, Perry, proclaimed that "[NATO's] main security front has swung away from Central Europe to its Southern Flank." He urged that future ministerial meetings should consider how NATO should engage the "responsible" states in North Africa in "a security dialogue and relationship," which would include the extension of existing bilateral contacts between North African countries and NATO.[104]

By September 1994, therefore, it would seem that a consensus had been virtually reached within the Atlantic Alliance concerning the launching of some form of initiative with the Mediterranean. Agreement would have to be obtained, though, about which non-NATO Mediterranean states should be invited by NATO to enter into a form of dialogue with the Atlantic Alliance. Here, work was going on behind the scenes under the auspices of NATO's Ad Hoc Group on the Mediterranean. This group reported to NATO's Permanent Representatives at NATO and also to the NATO defense ministers at Seville.[105]

In early December 1994 the NAC meeting in Brussels announced that as agreed at Istanbul, proposals had been examined with regard to promoting a dialogue, and now NATO was ready to establish contacts "on a case-by-case basis" with Mediterranean nonmember countries in order to strengthen regional stability. The Council in Permanent Session was directed to continue to review the situation and develop details of the dialogue and initiate appropriate contacts.[106] Finally, on February 8, 1995, an official statement was released that announced the beginning of NATO's Mediterranean Initiative. A "direct dialogue" would commence with Egypt, Israel, Mauritania, Morocco, and Tunisia. Further details of this official statement are examined in detail in the next chapter.

However, NATO's Mediterranean Initiative was not launched in the most favorable circumstances. It has been previously noted that on February 5, at a conference in Munich, US Secretary of Defense Perry had openly stated that North Africa posed a security threat to NATO. Arab governments and peoples would become even more outraged. In an interview with the German newspaper *Süddeutsche Zeitung,* that was timed to coincide with the opening of the security conference at Munich, then NATO secretary-general Willy Claes declared that Islamic fundamentalism was "at least as dangerous as Communism was," and that it would be impossible to reconcile Islamic fundamentalism and democracy.[107] One week later Claes moderated his remarks by dropping the comparison with Communism, but he maintained that Islamic fundamentalism was a major threat to NATO.

It was reported that Spanish diplomats lodged complaints against the NATO secretary-general. The Palestinian Pan-Arab daily, *al-Quds al-Arabi,* wrote that Claes was seeking to interfere in the internal affairs of Arab states. The newspaper continued: "Nobody can agree to swallow an insult such as this which has been directed at our own security and military establishments, which these Europeans think that they can use to advance their goals in whatever manner they please."[108] Even some liberal and secular commentators in the Arab world looked on NATO's Mediterranean Initiative as confrontational and an intended declaration of enmity against Islam and certain Arab-Islamic countries.[109] Clearly, the comments of the NATO secretary-general were highly insensitive and particularly untimely. It does not appear sensible to refer to threats stemming from a certain group at the very moment when one is seeking to begin a constructive dialogue with this same group. Furthermore, references to Islam should be careful not to label automatically all forms of Islam as "fundamentalist," because of the association—albeit mistaken—in the West of fundamentalism with terrorism and violence. Prior to the remarks of Claes, Arab publics were already highly suspicious of the motives and intentions of Western governments. The statement of the NATO secretary-general appeared to confirm their worst fears. And, in spite of what seems to have been months of painstaking planning concerning NATO's Mediterranean Initiative, the remarks of Claes had cast an ominous shadow over relations between NATO and the Mediterranean even before the dialogue had commenced.

CONCLUSION

The southern European members of NATO, and, in particular, France, Italy, and Spain, were an important driving force in promoting various trans-Mediterranean cooperation initiatives that culminated in the announcement of NATO's Mediterranean dialogue. Certain non-NATO Mediterranean states, especially Egypt, Malta, and Morocco, had lobbied for the start of a dialogue dealing with security issues in the broadened sense between states north and south of the Mediterranean. The first initiatives that were launched after the end of the Cold War were unsuccessful because of the unfavorable circumstances in the Maghreb and in the Middle East. Positive developments in the Middle East and the tacit agreement, as it were, to exclude the rogue state Libya from any dialogue prompted a new wave of trans-Mediterranean cooperation initiatives or resulted in the strengthening of ongoing north-south dialogues. Algeria remained a somewhat problematic case because of its serious internal civil unrest, although unlike other institutions and forums only NATO would decide not to enter into a dialogue with the government in Algiers.

Earlier in this study it was noted that in order for a constructive dialogue to commence and in order for CBMs and CSBMs to be successful, certain preconditions must be in place. One might argue that in the Mediterranean region as a whole, the circumstances were only favorable by 1993–94. Considerable progress had been made in the Middle East peace talks, and the EU had made it clear that it was determined to expend more resources on dealing with the political, economic, and social problems in the Maghreb in particular. But even in this period the military CBMs that ACRS was attempting to implement would not be implemented due to problems in Arab-Israeli relations with regard to WMD. It would seem that security issues in the broadest sense—political, economic, and societal, for example—would need to begin to be addressed before real progress could be made with regard to military CBMs and then CSBMs in the region. The WEU-Mediterranean dialogue with its exclusive focus on military and defense-related issues had achieved little by spring 1995.

NATO officials could take heed of the mistakes and failures of earlier dialogues and trans-Mediterranean cooperation initiatives. The Alliance's Strategic Concept had referred to an enlarged concept of security for the post–Cold War era and within this context had spoken of the value of dialogue, conflict prevention, and crisis management. But it was far from clear just what role NATO could play in the Mediterranean given that NATO was regarded by almost all people—especially in the southern and eastern Mediterranean—as a purely military organization. Some Atlantic Alliance officials including the secretary-general were still focusing on what they perceived to be potential if not actual threats rather than risks and challenges from the south. What would happen in practice is that NATO would attempt to reach out to Arab governments and publics, in particular, in an attempt to dispel their suspicions of the Atlantic Alliance. This would be no easy task. However, through a dialogue and a package of practical programs of cooperation, and through attempts to coordinate cooperation initiatives and dialogues with the activities of other trans-Mediterranean, NATO has endeavored to play its part in improving relations between governments and peoples north and south of the Mediterranean.

NOTES

[1] Ian O. Lesser, *Mediterranean Security: New Perspectives and Implications for US Policy* (Santa Monica, Ca.: RAND, 1992), 31.

[2] Address given by Manfred Wörner, Secretary-General of NATO, to the 36th Annual Session of the NAA, London, November 29, 1990, in *Change and Continuity in the North Atlantic Alliance—Speeches by the Secretary-General of NATO, Manfred Wörner* (Brussels: NATO Office of Information and Press, 1990), 262.

[3] Ali Hillal Dessouki, "The Impact on Relations between the Islamic World and Western Europe," in *The Implications of the Yugoslav Crisis for Western Europe's Foreign Relations,* Matthias Jopp, ed. (Paris: Chaillot Papers 17, Institute for Security Studies, WEU, 1994), 83–89.

[4] *Assembly of Western European Union, Proceedings, 42nd Session, December 1996, I Assembly Documents* (Paris: WEU), Doc.1543, November 4, 1996, "Security in the Mediterranean Region." Report submitted on behalf of the Political Committee by Mr. de Lipkowski, Rapporteur, 166.

[5] Carlos Echeverria, *Cooperation in Peacekeeping among the Euro-Mediterranean Armed Forces* (Paris: Chaillot Papers 35, Institute for Security Studies, WEU, 1999), 12–24.

[6] Joel Peters, *Pathways to Peace: The Multilateral Arab-Israeli Peace Talks* (London: The Royal Institute of International Affairs, 1996), 5–6.

[7] Joel Peters, "The Multilateral Dimension of the Middle East Peace Process," in *Mediterranean Politics, Vol.2,* Richard Gillespie, ed. (London: Pinter, 1996), 31–33.

[8] Bruce Jentleson, "The Middle East Arms Control and Regional Security (ACRS) Talks: Progress, Problems and Prospects," *Institute on Global Conflict and Cooperation, University of California, Policy Paper,* 26 (September 1996), 8.

[9] Fred Tanner, "The Euro-Med Partnership: Prospects for Arms Limitations and Confidence-Building after Malta," *The International Spectator* 32, 2 (April–June 1997): 10–11; and Ariel E. Levite and Emily B. Landau, "Confidence- and Security-Building Measures in the Middle East," in *Regional Security in the Middle East: Past, Present and Future,* Zeev Maoz, ed. (London and Portland, Oreg.: Frank Cass, 1997), 162–63.

[10] Peter M. Jones, "New Directions in Middle East Deterrence: Implications for Arms Control," *Middle East Review of International Affairs e-mail Journal,* 4, 4 (December 1997).

¹¹Jentleson, "The Middle East Arms Control and Regional Security (ACRS) Talks," 10–11.

¹²Speech by the secretary-general of NATO, Manfred Wörner, at the 1992 Sea Link Symposium, Annapolis, Maryland, June 18,1992.

¹³Final Communiqué of the Ministerial Meeting of the NAC, Copenhagen, June 6–7, 1991, in *NATO Review* 39, 3 (June 1991): paragraph 10, 32–33.

¹⁴*America and Europe: The Future of NATO and the Transatlantic Relationship.* Final Report of the NAA Presidential Task Force, Co-Chairmen Loic Bouvard (France) and Charlie Rose (United States) (Brussels: International Secretariat, 1993), 8, 6.

¹⁵*London Declaration on a Transformed North Atlantic Alliance—Issued by the Heads of State and Government Participating in the Meeting of the North Atlantic Council in London, 5–6 July 1990* (Brussels: NATO Office of Information and Press, 1990), paragraph 2.

¹⁶*The Alliance's Strategic Concept—Agreed by the Heads of State and Government Participating in the Meeting of the North Atlantic Council in Rome, 7–8 November 1991* (Brussels: NATO Office of Information and Press, 1991), paragraph 25.

¹⁷Mark Stenhouse and Bruce George, *NATO and Mediterranean Security: The New Central Region* (London: London Defence Studies, 22, Brassey's for The Centre for Defence Studies, London, 1994), 45–46.

¹⁸*Alliance Policy Framework on Weapons of Mass Destruction—Issued at the Ministerial Meeting of the North Atlantic Council in Istanbul, 9 June 1994* (Brussels: NATO Office of Information and Press, 1994).

¹⁹Ashton B. Carter and David B. Ormond, "Countering the Proliferation Risks: Adapting the Alliance to the New Security Environment," *NATO Review* 44, 5 (September 1996): 12–14.

²⁰NAA, Mediterranean Special Group, *Draft Interim Report—NATO's Role in the Mediterranean,* AP 115 GSM (97) 5 by Mr. Pedro Moya (Spain), Rapporteur (Brussels: International Secretariat, April 1997), 2.

²¹Roberto Aliboni, "Southern European Security: Perceptions and Problems," in *Southern European Security in the 1990s,* Roberto Aliboni, ed. (London and New York: Pinter, 1992), 2–3.

²²Ibid., 8–9.

²³Herminio Santos, "The Portuguese National Security Policy," in *Southern European Security in the 1990s,* 87–88, 92; and Fernanda Faria, "The Mediterranean: A New Priority in Portuguese Foreign Policy," *Mediterranean Politics* 1, 2 (autumn 1996): 212–30.

²⁴Willem van Eekelen, "European Security Interests in the Mediterranean," in *Report on the Sixth International Antalya Conference on Security and Cooperation,* organized by the Atlantic Council of Turkey, Antalya, November 2–5, 1995, 54.

[25]Stephen C. Calleya, "Malta's Post–Cold War Perspective on Mediterranean Security," in *Mediterranean Politics, Vol.1,* Richard Gillespie, ed. (London: Pinter, 1994), 138.

[26]CSCE, *Helsinki Final Act* (Helsinki: 1975), 111–12.

[27]Victor-Yves Ghebali, "Toward a Mediterranean Helsinki-Type Process," *Mediterranean Quarterly* 4, 1 (winter 1993): 92.

[28]Claire Spencer, "Building Confidence in the Mediterranean," *Mediterranean Politics* 2, 2 (autumn 1997): 32.

[29]Ghebali, "Toward a Mediterranean Helsinki-Type Process," 93.

[30]Fernando Rodrigo, "The End of the Reluctant Partner: Spain and Western Security in the 1990s," in *Southern European Security in the 1990s,* 112.

[31]Thanos Veremis, "European Political Cooperation and the Pursuit of Security: Toward a Southern Position?" in ibid., 31.

[32]Antonio Badini, "Efforts at Mediterranean Cooperation," in *Maelstrom: The United States, Southern Europe and the Challenges of the Mediterranean,* John W. Holmes, ed. (Cambridge, Mass.: The World Peace Foundation, 1995), 112–13.

[33]For the text of the Joint Document by France, Italy, Portugal, and Spain for a Conference on Security and Cooperation in the Mediterranean, see NAA, Civilian Affairs Committee, Sub-Committee on the Mediterranean Basin, *Interim Report,* AI 236 CC/MB (91) 6 by Mr. Augusto Borderas (Spain), Rapporteur (Brussels: International Secretariat, October 1991), Annex 2, 20–21.

[34]CSCE, *Report of the Meeting on the Mediterranean of the Conference on Security and Cooperation in Europe, foreseen by the Concluding Document of the Vienna Meeting 1986–Palma de Mallorca, 19 October 1990* (Palma de Mallorca, 1990), 14–25.

[35]Faria, "The Mediterranean: A New Priority in Portuguese Foreign Policy," 224–25.

[36]Christophe Carle, "France, the Mediterranean and Southern Europe Security," in *Southern European Security in the 1990s,* 48–49.

[37]Ghebali, "Toward a Mediterranean Helsinki-Type Process," 96–97; and Badini, "Efforts at Mediterranean Cooperation," 112–13.

[38]These points were noted by Professor Stefano Silvestri in an NAA Workshop on Confidence-Building Measures in the Mediterranean Region, Naples, April 11–12, 1997.

[39]Stephen C. Calleya, "Post–Cold War Regional Dynamics in the Mediterranean Area," *Mediterranean Quarterly* 7, 3 (summer 1996): 44.

[40]Tim Niblock, "North-South Socio-Economic Relations in the Mediterranean," in *Security Challenges in the Mediterranean Region,* Roberto Aliboni, George Joffé, and Tim Niblock, eds. (London and Portland, Oreg.: Frank Cass, 1996), 124.

⁴¹CSCE, *Helsinki Document—The Challenges of Change, July 1992* (Helsinki: 1992), paragraph 38; Chapter 10—The Mediterranean.

⁴²Ghebali, "Toward a Mediterranean Helsinki-Type Process," 99–101.

⁴³*Assembly of Western European Union, Doc. 1485, 6 November 1995,* "Parliamentary Co-operation in the Mediterranean." Report submitted on behalf of the Committee for Parliamentary and Public Relations by Mr. Kitsonis, Rapporteur, 14.

⁴⁴Carle, "France, the Mediterranean, and Southern European Security," 43; and Sarah Collinson, *Shore to Shore—The Politics of Migration in Euro-Maghreb Relations* (London, and Washington, D.C.: The Royal Institute of International Affairs and The Brookings Institution, 1996), 48.

⁴⁵Collinson, *Shore to Shore,* 48–49.

⁴⁶NAA, Borderas Report, October 1991, 14–15.

⁴⁷NAA, Civilian Affairs Committee, Sub-Committee on the Mediterranean Basin, *Report—Frameworks for Cooperation in the Mediterranean,* AM 259 CC/MB (95) 7 by Mr. Pedro Moya (Spain), Rapporteur (Brussels: International Secretariat, October 1995), 6.

⁴⁸Ghebali, "Toward a Mediterranean Helsinki-Type Process," 98–99; and Collinson, *Shore to Shore,* 50.

⁴⁹NAA, Civilian Affairs Committee, Sub-Committee on the Mediterranean Basin, *Interim Report,* AJ 236 CC/MB (92) 4 by Mr. Augusto Borderas (Spain), Rapporteur (Brussels: International Secretariat, November 1992), 8.

⁵⁰NAA, Political Committee, Sub-Committee on the Southern Region, *Draft Interim Report,* AM 106 PC/SR (95) 1 by Mr. Rodrigo de Rato (Spain), Rapporteur (Brussels: International Secretariat, May 1995), 4.

⁵¹Ibid., 7.

⁵²Margaret Blunden, "Insecurity on Europe's Southern Flank," *Survival* 36, 2 (summer 1994): 138.

⁵³*The Alliance's Strategic Concept.*

⁵⁴Ibid.

⁵⁵Speech by the secretary-general of NATO, Manfred Wörner, to the NAA, Madrid, October 21, 1991.

⁵⁶Speech by the secretary-general of NATO, Manfred Wörner, at the Alastair Buchan Memorial Lecture, London, November 23, 1988.

⁵⁷*The Alliance's Strategic Concept.*

⁵⁸Ibid.

⁵⁹Ibid.

⁶⁰NAA, Political Committee, *Europe and Transatlantic Security in a Revolutionary Age,* AK 244 PC (93) 6 by Mr. Bruce George (UK), General Rapporteur (Brussels: International Secretariat, October 1993), 9.

[61]Final Communiqué issued at the Ministerial Meeting of the NAC, Athens, June 10, 1993, in *NATO Review* 41, 3 (June 1993): paragraph 4, 31.

[62]NAA, Political Committee, Sub-Committee on the Southern Region, *Cooperation and Security in the Mediterranean,* AL 223 PC/SR (94) 2 by Mr. Rodrigo de Rato (Spain), Rapporteur (Brussels: International Secretariat, November 1994), 2, 10–11.

[63]Roberto Aliboni, "Collective Political Cooperation in the Mediterranean," in *Security Challenges in the Mediterranean Region,* 55.

[64]NAA, Moya Report, October 1995, 8–9.

[65]Secretariat of the CSCE, Prague, *CSCE Communication, No 161,* Prague, May 26, 1993—Chairman—Summary of the CSCE Mediterranean Seminar, Valletta, May 17–21, 1993.

[66]NAA, de Rato Report, November 1994, 15.

[67]CSCE 23rd Meeting of the Committee of Senior Officials, Prague, September 23, 1993, *23—CSO/Journal No.3,* Annex 4—Future relations between the nonparticipating Mediterranean states and the CSCE.

[68]CSCE 25th Meeting of the Committee of Senior Officials, Prague, March 3, 1994, *25—CSO/Journal No.2,* 3–4.

[69]NAA, de Rato Report, May 1995, 10.

[70]CSCE, *Budapest Document, December 1994—Budapest Summit Declaration—Towards a Common Partnership in a New Era* (Budapest, 1994), Section X, Mediterranean, 44–45.

[71]Stephen C. Calleya, *Navigating Regional Dynamics in the Post–Cold War World: Patterns of Relations in the Mediterranean Area* (Aldershot, and Brookfield, Vt.: Dartmouth, 1997), 215.

[72]Ibid., 12–13.

[73]WEU, Kitsonis Report, November 1995, 13.

[74]"Platform on European Security Interest," issued by WEU foreign ministers and defense ministers, The Hague, October 27, 1987, in *Western European Union: The Reactivation of WEU: Statements and Communiqués 1984 to 1987* (London: Secretariat-General of WEU, 1988), paragraph 4, 43.

[75]*Treaty on European Union* (Luxembourg: Office for Official Publications of the European Communities, 1992), Art J.4, paragraph 2, 126.

[76]"Declaration of the Member States of Western European Union which are also Members of the European Union on the Role of the Western European Union and its Relations with the European Union and the Atlantic Alliance," in *WEU Related Texts adopted at EC Summit, Maastricht—10 December 1991* (London: WEU Press and Information Section, December 10, 1991).

[77]WEU, *Council of Ministers, Petersberg Declaration, Bonn, 19 June 1992,* Part 2, "On Strengthening WEU's Operational Role," paragraph 4.

[78]*Assembly of Western European Union, Doc. 1468,* 12 June 1995, "European Armed Forces." Report submitted on behalf of the Defence Committee by Mr. de Decker, Rapporteur.

[79]WEU, *Petersberg Declaration,* Part 1—"On WEU and European Security," section 1, paragraph 18.

[80]WEU, de Lipkowski Report, November 1996, 157.

[81]NAA, de Rato Report, May 1995, 8.

[82]*46th Annual Report of WEU Council to the Assembly of WEU (1 July–31 December 1994).*

[83]Alfred Tovias, "The EU's Mediterranean Policies under Pressure," in *Mediterranean Politics, Vol.2,* 13.

[84]Collinson, *Shore to Shore,* 67; and Andres Ortega, "Relations with the Maghreb,"in *Maelstrom,* 53.

[85]Esther Barbé, "The Barcelona Conference: Launching Pad of a Process," *Mediterranean Politics* 1, 1 (summer 1996): 26.

[86]Abdelwahab Biad, "Security and Co-operation in the Mediterranean: A Southern Viewpoint," in *Security Challenges in the Mediterranean Region,* 46–47.

[87]NAA, Moya Report, October 1995, 3.

[88]NAA, de Rato Report, May 1995, 9.

[89]Ibid.

[90]NAA, Standing Committee, *Restructuring NAA Mediterranean Activities,* AN 33 SC (96) 9 by Mr. Pedro Moya (Spain), Rapporteur of the Sub-Committee on the Mediterranean Basin, and Mr. Anders Sjaastad (Norway), Chairman of the Defence and Security Committee (Brussels: International Secretariat, March 1996).

[91]NAA, de Rato Report, November 1994, 17.

[92]NAA, Political Committee, *NATO and the New Arc of Crisis: Dialectics of Russian Foreign Policy,* AJ 260 PC (92) 5 by Mr. Bruce George (United Kingdom), Rapporteur (Brussels: International Secretariat, November 1992), 13.

[93]NAA, Political Committee, Sub-Committee on the Southern Region, *Interim Report,* AJ 262 PC/SR (92) 4 by Mr. Miguel Herrero (Spain), Rapporteur (Brussels: International Secretariat, November 1992), 9.

[94]*America and Europe: The Future of NATO and the Transatlantic Relationship,* 21.

[95]NAA, de Rato Report, May 1995, 1.

[96]NAA, Moya Report, October 1995, 14–15.

[97]Stanley Sloan, "US Perspectives on NATO's Future," *International Affairs* 31, 2 (April 1995): 229–230.

[98]Final Communiqué of the Ministerial Meeting of the NAC, Athens, June 10, 1993, in *NATO Review* 41, 3 (June 1993): paragraph 11, 33.

[99] Press Communiqué M-1 (94) 3, Declaration of the Heads of State and Government Participating in the Meeting of the NAC, Brussels, January 10–11, 1994, paragraph 22.

[100] NAA, Political Committee, *Continental Drift,* AL 221 PC (94) 5 by Mr. Bruce George (United Kingdom), Rapporteur (Brussels: International Secretariat, November 1994), 23.

[101] NAA, Moya Report, October 1995, 5.

[102] Press Communiqué M-NAC-1 (94) 46—Ministerial Meeting of the NAC, Istanbul, June 9,1994, paragraph 29.

[103] NAA, George Report, November 1994, 24.

[104] NAA, de Rato Report, November 1994, 16; May 1995, 3, 4.

[105] Editor's Notes at the end of the article by Sergio Balanzino, "Appendix—Security Challenges in the Mediterranean Region," in *Security Challenges in the Mediterranean Region,* n.2, 192.

[106] Press Communiqué M-NAC-2 (94) 116—Ministerial Meeting of the NAC, Brussels, December 1, 1994, paragraph 19.

[107] Roger Boyes, "Muslim Militancy Is Next Big Threat Says NATO Chief," *The Times,* February 3, 1995.

[108] *Middle East International,* March 3, 1995, 16; and Haifaa A. Jawad, "Islam and the West: How Fundamental Is the Threat?" *The RUSI Journal* 140, 4 (August 1995): 34–35.

[109] Volker Perthes, "Security Perceptions and Cooperation in the Middle East: The Political Dimension," *The International Spectator* 31, 4 (October–December 1996): 55.

CHAPTER 7

NATO's Mediterranean Initiative

INTRODUCTION

The NATO-Mediterranean dialogue commenced when there was still talk among leading Western officials of risks, challenges, and even threats to NATO members from the south. Quite clearly, then, in these circumstances it was not going to be easy to win over the confidence and trust of the Mediterranean countries that entered the dialogue. From the very outset, therefore, there were serious questions concerning just how far the dialogue could develop.[1]

NATO officials use the phrase "Mediterranean dialogue partners" when referring to the six states involved in NATO's Mediterranean Initiative. Evidently, originally, NATO members were reluctant to employ the term "partner" as this may have implied—wrongly—that the NATO-Mediterranean dialogue was somehow comparable to the PfP program with states of central and eastern Europe. In reality, there is no partnership between the Atlantic Alliance and the Mediterranean dialogue countries or "partners." The six non-NATO Mediterranean countries do not regard themselves as partners. Moreover, certainly with reference to the five Arab countries, they do not want to be considered partners. There is still a deep-rooted suspicion of the West and of NATO, in particular, in the Arab world, and Arab governments, thus, do not want to be perceived as working too closely with NATO.

This chapter will examine how the NATO-Mediterranean dialogue has evolved from its beginnings in February 1995 to early 1999. In line with the suggestions made at the beginning of this study, the nature of the parties involved in the dialogue, and the type and range of issues discussed, the level at which the dialogue is conducted and the procedures adopted will be analyzed. In the period in question the dialogue has developed in three separate phases. The initial, exploratory phase was between February and November 1995. Officials from NATO's International Staff held preliminary discussions with representatives from five non-NATO Mediterranean countries. In the second phase, between November 1995 and spring 1997, the dialogue was broadened with the addition of a sixth country, Jordan. There was also a deepening of the dialogue concerning the type and range of issues that were discussed. The scope of the dialogue was widened with the introduction of practical programs of cooperation. In the third phase, from spring 1997 to early 1999, there was a further deepening and widening of the dialogue. The dialogue has been elevated to a higher level and given a more visible political profile with discussions largely conducted by the newly established body, the MCG. The procedures have also been altered with the adoption of a "16+1" format. The practical programs of cooperation now include limited activities in the military field. It will be observed that the NATO-Mediterranean dialogue has continued to be strengthened in spite of serious difficulties in the Middle East peace process.

As has already been indicated, the nineteen members of NATO do not share identical viewpoints on all issues. NATO is not a monolithic organization. A so-called Club Med grouping of states within NATO—Italy, Portugal, and Spain—has been identified. These states are especially interested in expanding ties with Arab states to their immediate south. The United States has played a major role in the Mediterranean since 1945 and its views remain of paramount importance. Canada and the northern European members of NATO appear to be less enthusiastic about the Atlantic Alliance bolstering its ties with the Mediterranean dialogue countries. It will be seen that France, too, in spite of its traditional interest in the Mediterranean, is not keen on promoting the NATO-Mediterranean dialogue. The six Mediterranean dialogue countries each also have different perspectives on their relations with NATO. Given these circumstances, it is not surprising that the NATO-Mediterranean dialogue has only been able to evolve gradually and incrementally. But progress has been made as will be observed. How far the dialogue will continue to develop as a confidence-building process, and to what extent more military-type CBMs will be introduced and implemented within the practical programs of cooperation linked with the dialogue, remain open questions.

MEMBERSHIP OF THE NATO-MEDITERRANEAN DIALOGUE

As of early 1999 six Mediterranean dialogue countries—Egypt, Israel, Jordan, Mauritania, Morocco, and Tunisia—were engaged in discussions and participating in practical programs of cooperation with NATO. This looks like a motley grouping of states. In strictly geographic terms, Jordan and Mauritania are not Mediterranean countries. Perhaps what is most striking is the inclusion of Israel in the same grouping with five Arab countries. There is a mixture of states from the western and eastern shores of the Mediterranean. There is also a definite Middle East component. The six Mediterranean dialogue countries differ in the nature of their political systems and in the condition of their economies. As well as certain tensions between the Arab countries and Israel, there have also been problems in relations between Mauritania and Morocco.

On closer inspection, though, there are certain criteria that each of these states satisfies. Each was invited to join the dialogue after a consensus had been agreed on among NATO members. Each Mediterranean dialogue country accepted the invitation. Clearly, before invitations were formally extended, NATO officials had sounded out the state in question and had ensured that their invitation would not be rejected. It appears that Jordan was especially eager to be invited. In part, at least, this was probably because of its exclusion from the first wave of states invited to join the dialogue. Each of the six states has a reasonably stable government, unlike, for example, Algeria. The leadership of each of the six states has some claim to legitimacy. Mauritania was also included because of its geostrategic importance for Portugal and Spain. Each of the six states is, in general, pro-Western. The Mediterranean "rogue" states of Libya and Syria, and Syria's client state Lebanon, have not been invited to participate in the dialogue. They would probably have refused the offer. Each of the six states has normalized their relations with each other, including with Israel. However, with problems in the Middle East peace process following the formation of a Likud-led coalition government in 1996, some Arab states decided to freeze their relations with Israel in protest.

Initially, it seemed that only three Mediterranean states would be invited to participate in a NATO-Mediterranean dialogue. These states were Egypt, Morocco, and Tunisia. Apparently, as the United States gradually warmed to the idea of the dialogue, the Clinton administration strongly advocated that Israel also be invited. One may assume that the

United States insisted that the inclusion of Israel only proceed on the condition that there be no interference in the ongoing Middle East peace process. Concerned about the security of Madeira and the Canary Islands, Portugal and Spain promoted the cause of Mauritania. The unsubstantiated reports that Iraq might have deployed or attempted to deploy SCUD missiles in Mauritania in 1991 at the time of the Gulf Crisis might have made Portugal and Spain more eager to lobby for Mauritania. Once a member of the NATO-Mediterranean dialogue, the Mauritanian government would less likely be tempted by Iraqi offers of military weaponry. The governments in Lisbon and Madrid have fostered close relations with the administration in Nouakchott. Officials in Spain and Portugal would have been encouraged by the democratization process that commenced in Mauritania in 1992. As a consequence, by February 1995, what had originally been envisioned as the three had become the five when NATO's Mediterranean Initiative was formally announced.

At the very beginning of the NATO-Mediterranean dialogue, it was declared that other states could later join the dialogue:

> This is a progressive initiative and an extension of the dialogue to other Mediterranean countries, which are willing and able to contribute to the peace and security of the region, will be envisaged after the initial round of dialogue.[2]

The NAC Ministerial Meeting at Noordwijk at the end of May 1995 noted that it hoped that further discussions would lead to the establishment of a fruitful dialogue "with these and other Mediterranean countries."[3]

In November 1995 Jordan joined the NATO-Mediterranean dialogue. Jordanian officials may have argued that their country deserved to be invited because of its contribution to peace and security in the region, having earlier concluded a peace treaty with Israel. In January 1996 Crown Prince Hassan of Jordan, the highest ranking Arab leader to do so, visited NATO Headquarters in Brussels. Given its close interest in NATO's Mediterranean Initiative, it seems surprising that Jordan was not invited to join the dialogue in February 1995. Perhaps certain NATO members had held a lingering suspicion of Jordan, remembering that King Hussein had publicly supported Iraq in the 1990–91 Gulf crisis.

No other Mediterranean countries have been added to the NATO-Mediterranean dialogue. Algeria appears to have become a particularly controversial case. Algeria is included in other dialogues and initiatives concerning the Mediterranean and so it may seem that the exclusion of Algeria from NATO's Mediterranean Initiative is somewhat odd. Spain, with the backing of Italy and Portugal, had pressed for NATO to invite to talks all states involved in the WEU-Mediterranean dialogue; this would have included Algeria. It seems, however, that French officials were opposed to the admission of Algeria on the grounds that this would only complicate an already complex situation in Algeria. Also, it has been argued that the French authorities were not keen for NATO to assume an active involvement in Algeria, a country that had close political, economic, and military ties with France.[4] NATO allies were certainly divided over how to tackle the Algerian problem. The French were willing to maintain close ties with the military rulers while the United States was more in favor of entering into discussions with moderate Islamists active in Algeria. By early 1995, with the worsening domestic situation in Algeria it is understandable that NATO officials would not have wanted to take Algeria on board at the start of their Mediterranean Initiative. The Atlantic Alliance would not have wished to be perceived as giving support to the military-backed regime in Algiers. However, not only the Algerian authorities, but also some elements in the local media felt victimized by NATO's position. For example, it was reported that the NATO-Mediterranean dialogue was directed against Algeria and against "Islamization" in general. There were also complaints that NATO had placed Algeria in the same camp as Iran.[5]

Without a change in leadership and a dramatic turnabout in their foreign policies, it is exceedingly unlikely that Lebanon, Libya, or Syria would be invited to participate in the NATO-Mediterranean dialogue. As part of their interest in strengthening connections between Europe, the Mediterranean, and the Gulf, Jordanian officials have expressed a tentative desire to include Gulf states in the dialogue.[6] For many in NATO this would probably be perceived as stretching the definition of the Mediterranean too far. Most probably, NATO members would want to keep the NATO-Mediterranean dialogue separate from the turbulent Gulf area. Key NATO countries already have their own bilateral security agreements with various Gulf states.

A further broadening of the NATO-Mediterranean dialogue in terms of the admission of new Mediterranean dialogue countries is thus very unlikely for the foreseeable future. Bearing in mind the potential problems the broad-based CSCM may have encountered because of the diverse nature of its many anticipated members, and given the difficulties the Barcelona Process is experiencing between the EU and its twelve

Mediterranean partners (see next chapter), it is perhaps sensible for NATO officials to restrict the NATO-Mediterranean dialogue to a relatively small but more manageable number of states.

THE FIRST PHASE

It seems that officials in Italy and Spain were disappointed with the initial low-key nature of the dialogue. The Spanish authorities had pressed for the immediate inclusion of what would have been sensitive military issues. They were in favor of inviting to NATO headquarters military officers from the Mediterranean dialogue countries. Joint military exercises with the dialogue countries were also envisioned. Italian officials had argued that the dialogue should not only be conducted with governmental representatives from the Mediterranean dialogue countries. From the outset, for example, they had wanted to hold talks in workshops and seminars with journalists and academics. Originally, both Italy and Spain were in favor of the involvement of NATO's International Staff in the planning of the NATO-Mediterranean dialogue, and the participation of the more high level Political Committee of NATO in the actual discussions. An arrangement along these lines would only eventually emerge in the third phase of the NATO-Mediterranean dialogue.

The official announcement of the start of the dialogue in February 1995 had stated that this was a "direct dialogue" whose aims were "to contribute to security and stability in the Mediterranean as a whole, to achieve a better mutual understanding and to correct any misunderstandings of the Alliance's purposes that could lead to a perception of threat."[7] As previously noted, the speeches of leading Western officials, including one by NATO Secretary-General Claes, had contributed to the confusion by spreading the notion that the north was fearful of an imminent threat of some sort from the south. In turn, this had led many in the south to conclude that the north was using such arguments as a pretext to threaten to intervene in the internal affairs of Arab countries. Quite clearly, there was much need for mutual confidence-building. Contrary to the wishes, if not expectations, of officials from Italy and Spain, the NATO-Mediterranean dialogue would be conducted at a relatively low level. Talks were to be held at NATO Headquarters on a bilateral basis between officials from the Political Affairs Division of NATO's International Staff and representatives from the embassies in Brussels of the Mediterranean dialogue countries. However, NATO's International Staff would work on instructions from the NAC and NATO's Political Committee. In effect, five separate dialogues would commence between the International Staff and each of the Mediterranean dialogue countries. At this initial phase, there was no intention of multilateralizing the talks by encouraging the five Mediterranean countries to hold talks with NATO officials in a single group. Differences and divisions between the five Mediterranean dialogue countries, and especially between the Arab states and Israel precluded such an option. As the head of NATO Multinational and Regional Affairs in the Political Affairs Division at the time noted, any real multilateralization of the dialogue would make it much more vulnerable to political developments in the region.[8]

From the very beginning of the NATO-Mediterranean dialogue, however, there was speculative talk about what the dialogue might eventually lead to. Perhaps the most exaggerated of these speeches was the one made by Joseph Kruzel, then US deputy assistant secretary of defense for European and NATO Affairs, at AFSOUTH headquarters in Naples on February 27, 1995. In a declaration that was clearly not representative of the US official line, while referring to threats from the south, Kruzel spoke of the possibility of deepening cooperation with the dialogue countries. He referred to future regular visits of defense ministers and the involvement of military officers in seminars and exercises. In turn, Kruzel remarked, PfP-type arrangements for the Mediterranean could emerge. Even the possibility of eventual membership of NATO was mentioned![9] Obviously, Kruzel was a little overenthusiastic in his observations. Article 10 of the North Atlantic Treaty would have to be amended before any Mediterranean dialogue country could join the Atlantic Alliance. According to the article, "other European" states could be invited to accede to the treaty.

Interestingly, Kruzel had not imagined that a threat to NATO could stem from Mediterranean dialogue countries themselves. Presumably he had in mind rogue states such as Libya when referring to threats from the south. Apparently, Gadhafi reacted vehemently to the announcement of the NATO-Mediterranean dialogue. Reportedly, he called for the launching of a jihad should NATO attempt to expand its influence in Africa. The Libyan authorities warned ambassadors from Egypt, Mauritania, Morocco, and Tunisia that as a result of the launching of the NATO-Mediterranean dialogue, there was a danger of increased interference in the internal affairs of Arab states.[10]

The dialogue commenced with a number of courtesy calls on NATO Secretary-General Willy Claes. Starting on February 24, Claes began to receive individually the ambassadors of the Mediterranean dialogue countries based in Brussels.[11] The dialogue would become truly operational on May 15, when NATO's International Staff held their first meeting with one of the dialogue countries, in this case Mauritania. Why the three months delay? Technical reasons have been suggested. Or was it that the Mediterranean dialogue countries were hesitant to begin the dialogue because of the earlier remarks of Claes with regard to the Islamic threat? In the spring and summer of 1995 two rounds of bilateral talks were concluded between NATO's International Staff and four of the five dialogue countries. Morocco had refused to participate, in protest of the comments made by Claes. The first meetings between officials from NATO and Morocco would only take place after November 8, 1995, when the second phase of the dialogue had commenced.[12]

These first rounds of talks were simply exploratory in nature. NATO officials explained the nature and purpose of the Atlantic Alliance and sounded out the concerns of the Mediterranean dialogue countries. Taking into account how the dialogue countries responded, the NAC would then decide whether to proceed further with the dialogue. Nevertheless, it seems that there were attempts even at this initial phase to include issues of military significance in the discussions. For example, the NAC meeting in Noordwijk in May 1995 stated that NATO should inform the Mediterranean dialogue countries of the Atlantic Alliance's "new missions of peacekeeping under the authority of the UN or the responsibility of the OSCE."[13] Egypt, Jordan, and Morocco would later participate with NATO units in peacekeeping missions in Bosnia as part of IFOR and SFOR. Although this participation was not directly related to developments in the NATO-Mediterranean dialogue, nonetheless it could have future beneficial consequences for the possible strengthening of the dialogue.

In mid-1995, NATO's director for nuclear planning argued that the Atlantic Alliance should inform the Mediterranean dialogue countries about NATO's Policy Framework on the Proliferation of Weapons of Mass Destruction.[14] The NAC, however, has not given the MCG a mandate to discuss the proliferation issue with dialogue countries. NATO officials may only give briefings on this topic. They may not enter into a debate. Clearly, as noted previously, the proliferation issue is a highly sensitive one, and NATO officials do not want the NATO-Mediterranean dialogue to suffer the same fate as ACRS when a clash of opinions between Egypt and Israel on this topic led to the collapse of talks. It has also been suggested that the Arab countries in the dialogue might feel particularly aggrieved if, in the context of possible discussions with

NATO on the proliferation issue, they believe that the Atlantic Alliance is attempting to divest them of particular weaponry. The military balance in the Mediterranean region already weighs heavily in favor of states to the north.[15]

In practice, therefore, sensitive military issues were not discussed in the initial phase of the NATO-Mediterranean dialogue. NATO officials were aware that much work needed to be done to build confidence and trust with the Mediterranean dialogue countries. In mid-October 1995, speaking at a seminar organized by the RAND Corporation, the deputy secretary-general of NATO, Sergio Balanzino, proclaimed that the initiative "still has some way to go" before the participants reach a degree of understanding satisfactory to him.[16]

In spite of some difficulties, enough progress had been made for the NAC to decide on November 8, 1995, to pursue the dialogue further by entering into what amounted to a new, second phase. The scope of NATO's Mediterranean Initiative would be widened. The Mediterranean dialogue countries would be provided an opportunity to participate in certain cooperative activities with NATO members. The membership of the dialogue was broadened. Jordan was invited to enter the talks. These decisions, for some reason, were only made public in a communiqué released by the NAC on December 5, 1995.[17] The cooperative activities with NATO members would be regarded by Atlantic Alliance officials as part of the NATO-Mediterranean dialogue.

THE SECOND PHASE

Interestingly, the communiqué of the NAC meeting in Brussels in December 1995 implied that the NATO-Mediterranean dialogue was still not fully established, as it stipulated that the NAC was exploring "the possibilities for a permanent dialogue with countries in the region."[18] However, by the end of the second phase of the dialogue in spring 1997 the talks between NATO officials and the Mediterranean-dialogue countries were clearly based on a much more established footing.

Discussions continued to proceed on the basis—in theory at least—of two rounds of talks each year between NATO's International Staff and each of the Mediterranean dialogue countries. In this phase further sessions could be requested by any of the dialogue countries. But attendance at these sessions was not guaranteed. Apparently, for example, in 1996 some of the Mediterranean dialogue countries participated in only one session. One country did not hold any discussions with NATO's

International Staff in 1996. Nonattendance may have been in reaction to problems in the Middle East peace process, or, quite simply, it may have been due to a lack of interest or understanding of the dialogue on the part of the dialogue countries concerned. Talks were no longer merely exploratory. Topics discussed included political, social, and economic developments in the Mediterranean and the prospects for regional cooperation. More sensitive issues, particularly those related to military security matters were carefully avoided—for the same reasons that prevented an open discussion of the proliferation question. Thus, limits were placed on the type and range of issues that could be talked about.

A novel feature of this second phase of the dialogue was the introduction of an element of multilateralization. For instance, in June and December 1996 the six Mediterranean dialogue countries were invited to attend as a group briefings related to the recently concluded meetings of the NAC. Again, evidently, there were some absentees on each of these occasions. But a civil-emergency planning briefing organized by NATO officials in September 1996 was attended by all six dialogue countries. An unofficial Non-Paper on the NATO-Mediterranean dialogue produced by NATO at the end of 1996 recommended that further multilateral briefings be held on a regular basis after NAC meetings and other major events. It is important to note that only "briefings" were referred to here. There was no intention of holding discussions and debates with the six Mediterranean dialogue countries as a group. Israel and the Arab states in question did not want to be seen as a definite grouping. Discussions within the dialogue remained strictly bilateral. Both NATO officials and representatives of the dialogue countries favored continuing this arrangement.

More important, in this second phase the scope of the NATO-Mediterranean dialogue was widened through NATO offering the dialogue countries various cooperative activities. Contacts would no longer be solely with diplomatic officials. These cooperative activities covered the fields of information and science. With regard to information, NATO tentatively began to establish contacts with so-called opinion leaders from the dialogue countries in order to explain the aims and objectives of the Atlantic Alliance. It appears that NATO officials were considering initiating and developing links with certain groups such as academics, parliamentarians, journalists, and religious leaders. In October and November 1996 two conferences were held in Rome devoted to security issues in the Mediterranean to which academics from the dialogue countries were invited. In November 1996 certain "opinion leaders" from these countries were received at NATO Headquarters. It appears, though, that it was eventually decided not to invite representatives from the media. Hosting journalists from the Mediterranean dialogue countries would entail certain risks. Which reporters from which newspapers should be invited? And there would be no guarantee that these journalists would write favorable reviews on NATO for their domestic audiences.

In the field of science, NATO officials were able to exploit the work of the Science Committee and the Committee on the Challenges of Modern Society (CCMS). These are products of NATO's so-called Third Dimension. The Third Dimension was developed in the late 1950s with the purpose of demonstrating that NATO was not simply a political and military organization. In November 1995 the Mediterranean dialogue countries were requested to nominate certain "contact points" to receive and disseminate information relating to NATO's scientific activities. In the same month, NATO decided that the dialogue countries could send their scientists on a self-funded basis to attend scientific meetings organized by the Science Committee. In December 1996 it was agreed that the dialogue countries could work together with the CCMS on certain pilot studies. However, in early 1997 NATO officials were expressing their disappointment at the lack of interest shown by the dialogue countries in NATO's scientific work. It seems that in 1996 the dialogue countries produced no concrete proposals concerning possible joint scientific work with NATO. This lack of interest was all the more surprising as apparently considerable sums of money were potentially available for cooperative work in certain fields. In contrast, in this period there was a serious shortage of money to fund cooperative activities in the information sphere. The program of activities in information was basically self-funding. Mediterranean dialogue countries complained of the expenses involved in participating in such activities.

The unofficial Non-Paper recommended the expansion of cooperative activities in the fields of information and science. Moreover, significantly, the Non-Paper also proposed the introduction of cooperative activities in the military sphere. The first briefing on civil-emergency planning had been an initial step toward developing civil-military cooperation. The Non-Paper suggested that courses on civil-military cooperation should be offered to representatives from the Mediterranean dialogue countries. More ambitiously, the Non-Paper proposed that other cooperative activities in the military domain should be considered. Courses and sessions on peacekeeping, arms control, and verification could be organized. A workshop on data management in support of arms control was recommended with the proviso that this should not duplicate ongoing work in the Barcelona Process or interfere with current activities related to the Middle East peace process. Briefings on crisis management were also to be considered.

There was now a real prospect of military officers from the Mediterranean dialogue countries attending various courses at the NATO school at Oberammegau or at the NATO Defense College in Rome. In 1996 the Defense College's commandant for curriculum planning visited defense colleges in Egypt, Israel, Jordan, and Tunisia to explore areas of possible cooperation.

In November 1996 NATO's Council in Permanent Session conducted a brainstorming session to examine how the NATO-Mediterranean dialogue could be further pursued. The following month the NAC declared that the dialogue would be enhanced in a "progressive way." The NAC tasked the Council in Permanent Session to report at the next NAC meeting on the implementation of activities foreseen in an earlier report prepared by the same Council.[19] As a result, the decisions taken at the Sintra meeting of the NAC in spring 1997 heralded the commencement of a third phase of the dialogue.

By the end of 1996 the dialogue was well established. However, there were still a number of problems. For example, the representatives of certain Arab countries complained that Israeli military officials were allowed to participate in the dialogue. Arab officials also protested that NATO was not informing them of the content of the dialogue between the International Staff and Israel. There was a call for more "transparency" in the NATO-Mediterranean dialogue. Quite clearly, more work still needed to be done to build further confidence and trust.

THE THIRD PHASE

In this phase, at the time of writing in early 1999, no new Mediterranean countries had been added to the dialogue. The scope of the dialogue had widened, though, to include more cooperative activities, particularly in the military field. Also, significantly, the dialogue was elevated to a higher level with the formation of the MCG. As a consequence, there were alterations in the procedures involved in the conduct of the dialogue.

In May 1997, at Sintra, the NAC declared: "We want to further enhance this dialogue and improve its overall political visibility as an effort of confidence-building and cooperation that contributes to stability." The ministers recommended that the upcoming meeting of heads of state and government in Madrid in July 1997 should establish under the authority of the NAC a new committee that would have overall responsibility for the Mediterranean dialogue.[20] This was the first occasion an official NATO text referred to the NATO-Mediterranean dialogue in terms of "confidence-building." The Madrid Declaration of the NATO heads of state and government would again mention confidence-building. In line with the recommendations made at Sintra, the Madrid meeting agreed to form the MCG with a mandate to manage the NATO-Mediterranean dialogue.[21]

The MCG replaced the Ad Hoc Group on the Mediterranean. The Expert Working Group on the Middle East and the Maghreb continued to function. The chairperson of NATO's Political Committee could also chair meetings of the MCG. The MCG would normally meet at the level of political advisers from each of the sixteen national delegations at NATO headquarters. There was also the possibility of holding "reinforced" meetings with representatives from the capitals of the NATO member states. Clearly, the MCG was a much more politically influential body than its predecessor, the Ad Hoc Group on the Mediterranean. The MCG could meet as a group of sixteen. It would also gather to meet separately once every twelve months with representatives of the Mediterranean dialogue countries in a "16+1" format. If necessary, additional ad hoc "16+1" meetings could also be convened.[22] The International Staff of NATO could also continue to meet officials from the six dialogue countries on an individual basis, although now on a once-a-year basis. The first "16+1" meetings were held on November 20–21, 1997.

According to the NAA Rapporteur, Pedro Moya, the establishment of the MCG raised the level of the NATO-Mediterranean dialogue from one at an administrative level to one that had a much more visible political profile: "This is, therefore, a significant qualitative development in Mediterranean partnership." With the formation of the MCG, according to Moya, Mediterranean issues would become a permanent item on the agenda of the Political Committee. Moya believed that this should enable a real Mediterranean policy to emerge progressively. Unlike the previous Ad Hoc Group on the Mediterranean, the MCG could make recommendations to the Political Committee and by extension to the NAC. In the words of Moya, dialogue countries were no longer restricted to contacts with NATO officials whose room for maneuver was limited, but were in direct touch with representatives from the national delegations at NATO Headquarters in Brussels.[23]

In practice, in spite of the creation of the MCG, NATO officials and representatives of the Mediterranean dialogue countries continued to discuss more or less the same issues that had been previously raised in the second phase of the dialogue with the International Staff of NATO. For example, there was still no mandate to talk in detail about the proliferation of WMD in the Mediterranean. Only briefings could be given on this issue. It seems that both NATO officials and the dialogue countries were not prepared to deepen the dialogue by tackling issues that were perceived to be more politically sensitive. More emphasis and importance was thus placed on the cooperative activities between NATO and dialogue countries.

By the end of 1997 the first of what seems to have become an annual Work Program of Cooperation had been prepared by the Atlantic Alliance. The program referred to cooperation in the fields of science, information, and civil-emergency planning, Significantly, limited cooperation in the military sphere was also now included. In the science field, a new feature was that experts from the dialogue countries could be invited as key speakers to participate in NATO-sponsored advanced research workshops or could be invited to attend lectures in NATO-sponsored advanced study institutes. Courses on civil-emergency planning and civil-military cooperation would include medical evacuation workshops and civil protection seminars. Another new feature was the linkage between these cooperative activities and the PfP program with central and eastern European states. For instance, in 1998 the six Mediterranean dialogue countries were invited to participate in a number of PfP activities concerning civil-emergency planning, although this would be on a self-funding basis.

In the information field, for the first time in October 1997 a group of high-level parliamentarians from the dialogue countries visited NATO Headquarters. NATO Secretary-General Solana referred to the meeting as an important contribution to confidence-building.[24] In November 1997 NATO helped sponsor two academic conferences in Rome that analyzed Mediterranean security issues. One of these was organized with the support of the RAND Corporation and focused specifically on the progress and prospects for the NATO-Mediterranean dialogue. In December 1997 another international seminar on security issues in the Mediterranean was convened in Ebenhausen, Germany, with NATO support. In 1998 NATO officials gave briefings in Brussels to academics and now journalists, and also to parliamentarians from the foreign affairs and defense committees of the Mediterranean dialogue countries. NATO also awarded its first Institutional Scholarships to scholars from the dialogue countries. Also, in May 1998 NATO officials agreed to establish contact points in the embassy of a NATO member state in each of the capitals of the Mediterranean dialogue countries. These contact points would be a source of information on NATO activities for the general public in the dialogue countries.[25]

A tentative start was made for cooperating in military activities. Potentially, there was much expertise that NATO could offer the Mediterranean dialogue countries. The Atlantic Alliance could also draw on the experience and make use of some of the resources available for the PfP programs. Between September and December 1997 courses on peacekeeping, military forces and environmental problems, European security cooperation and also on civil-emergency planning and civil-military cooperation were opened to Mediterranean dialogue countries at the NATO School at Oberammergau. In the same period a number of conferences, seminars, symposia, and visits were organized by the NATO Military Committee. Military officers from the dialogue countries could learn, for example, about the PfP work programs, maritime peace-support operations, air operations and humanitarian aid, maritime safety, mine warfare, and maritime counterterrorism. Further courses were held at Oberammergau in 1998 covering such issues as arms control and European security. In April 1998 the first course for military officers from the dialogue countries was opened at the NATO Defense School in Rome. Perhaps surprisingly, however, Morocco did not send officers. Only an observer from the Moroccan embassy in Rome participated. The aim of the course was for each participant to develop a mutual understanding of the others' security concerns.[26] A major drawback, however, was that most of these courses and activities were meant to operate on a self-funding basis. It seems, though, that in certain cases money could be provided to assist the Mediterranean dialogue countries.

Concerned not to stir up public opinion at home, it appears that the governments of the Mediterranean dialogue countries were slow to cooperate more extensively with NATO in the military field. Eventually, however, in 1998, military officers from the dialogue countries agreed to observe PfP activities in the fields of search and rescue, maritime safety and medical evacuation, and exercises related to peace support and humanitarian relief. Port visits to the dialogue countries by STANAVFORMED were also planned.[27] Joint participation in military exercises, though, is unlikely for the foreseeable future. This is in spite of the involvement of units from Egypt, Jordan, and Morocco in peacekeeping operations in Bosnia. In these operations the armed forces of the three Arab countries were adopting NATO operational procedures and familiarizing themselves with rules of engagement and the practice of command and control. Close personal links were being established with NATO officers. It should also be remembered that the armed forces of the Mediterranean dialogue countries have already been involved in bilateral and multilateral programs of military cooperation with various NATO member states. It appears that cooperating in the military field with one or several NATO member states is far less problematic politically for Arab governments than coordinating activities with NATO as an organization. The suspicion remains among Arabs generally that NATO is seeking to interfere in their internal affairs. NATO officials, therefore, will have to proceed especially carefully when attempting to explain to dialogue countries the role of NATO in crisis management. Arab officials and Arab publics alike suspect that NATO intends to "manage" future crises in their part of the world as a pretext for the Atlantic Alliance's possible expansionist aims.

For information and science activities, and for joint work associated with civil-emergency planning, up to three times more funding was made available for 1998 in comparison to 1997. But this was still an insignificant amount compared to the expenditure for the PfP programs. The Mediterranean dialogue countries were fully aware of this situation and the Arab states have complained that this was evidence that NATO was not taking the Mediterranean Initiative seriously enough.

As the dialogue has progressed it appears that each Mediterranean dialogue country has certain particular issues they wish to discuss with NATO officials. The Atlantic Alliance is engaged in six different dialogues in the "16+1" format. Each dialogue country has also chosen to participate or not in various cooperative activities. One may presume that a shortage of funds has prevented certain states from participating in particular activities. It would seem, for example, that Israel has the resources and expertise to join more cooperative activities than Mauritania. One may also safely assume that the nature of discussions between Israeli representatives and NATO officials differs in substance and content from talks between the MCG and Mauritania. Speaking in Rome in November 1997, Secretary-General Solana emphasized that the NATO-Mediterranean dialogue was "a dialogue of variable geometry." Mediterranean countries would be able to structure the dialogue according to their particular needs.[28] It is not clear, though, how far NATO officials are prepared to allow each of the six separate dialogues to diverge.

NATO officials have also argued that the Mediterranean Initiative is based on the principle of nondiscrimination, and that "what is offered to one dialogue partner is offered to all the others in the dialogue."[29] As the separate dialogues presumably continue to further diverge, it will becoming increasingly difficult if not impossible to offer the same to all dialogue countries. And in that event, accusations of discrimination in favor of one dialogue country over another are more likely to ensue.

There is certainly a need to build more confidence between NATO and the Arab members of the Mediterranean dialogue. More information and transparency is required. The so-called *El Mundo* incident is a striking case in point. The meeting of the NAC in Defense Ministers session in Brussels on December 2, 1997, addressed the issue of the new command structures for NATO in the Mediterranean and elsewhere. Three days later an article appeared in the Spanish newspaper *El Mundo*. The report referred to the role of Spanish armed forces in possible future military operations in the Maghreb "in support of peace." The article stated that NATO could deploy an army corps of between twenty and fifty thousand troops in North Africa.[30] The publication of this report triggered a wave of protests in the Maghreb stirred up by Gadhafi's accusations that a NATO attack on North Africa was imminent. Spanish and other NATO officials immediately sought to stabilize the situation by explaining that the newspaper report had misinterpreted the results of the NAC meeting. However, several weeks after the incident Arab officials were still raising the issue and speculating about what contingency plans NATO actually possessed concerning a possible armed intervention in North Africa.

THE PERSPECTIVES OF NATO STATES

The keen interest of France, Italy, and Spain concerning various initiatives and dialogues with regard to the Mediterranean has been discussed previously. Italy and Spain were the main sponsors of the NATO-Mediterranean dialogue. Interestingly, Javier Solana was Spanish foreign minister when Spain was forcefully advocating the launching of a NATO dialogue with the Mediterranean. Spanish officials had also lobbied for the creation of the MCG. Portugal has assumed a greater interest in the progress of NATO's Mediterranean Initiative. However, French officials have been much less enthusiastic about the NATO-Mediterranean dialogue.

Picking up on an idea that had been earlier raised by the American official Joseph Kruzel in October 1995 at an informal meeting of NATO defense ministers in Williamsburg, Italian Defense Minister General Domenico Corcione spoke in favor of creating a PfP-type arrangement for the Mediterranean. Corcione envisioned here Mediterranean dialogue countries sending observers to NATO military exercises and planning joint operations with NATO in politically sensitive fields such as search and relief operations. According to Corcione, cooperation in the military field would have to develop incrementally through, initially, reciprocal visits, participation in courses and seminars, and information exchange.[31] Obviously, in the third phase of the NATO-Mediterranean dialogue the first steps have been taken toward possibly developing cooperation in the military field. However, Italian officials have backtracked on the idea of a PfP for the Mediterranean that could somehow be comparable to the PfP programs for central and eastern Europe. The Italians appear to have become more aware that the term "partnership" with regard to the Mediterranean is one that many NATO members are instinctively opposed to. There is a feeling among most NATO states that the Mediterranean should not appear to be elevated to the same level of importance as NATO's relations with central and eastern Europe and with Russia.

Portugal, though, has continued to lobby for the eventual adoption of a "PfP-inspired model" for the Mediterranean. This model was forcefully advocated by Ambassador Antonio Monteiro, the deputy director of the Multilateral Affairs Department of the Portuguese Foreign Ministry, at an international conference held in Lisbon in December 1996. Monteiro noted that different packages could eventually be prepared for each Mediterranean dialogue country according to the principle of "variable geometry." In this PfP-inspired model, cooperation in the fields of civil-emergency planning, crisis management, the civilian control of armed forces and defense structures, defense planning and budgeting, defense policy and strategy, military education, training and doctrine, and peacekeeping-related exercises and training could be included. Monteiro argued that beforehand intermediate measures could be taken to enhance the dialogue. In this context, he suggested that dialogue countries be allowed to participate in NAC meetings with observer status, to establish observer missions at NATO Headquarters, to attend seminars that focused on the work of IFOR in the Balkans, and to assist in the planning of NATO military exercises and also participate in these exercises.[32]

It would seem that Monteiro's proposals were far too ambitious. For example, it was exceedingly difficult to imagine that the armed forces of the Arab dialogue countries would readily agree to civilian control.

Evidently, Portugal had even at one time advocated an actual Partnership for the Mediterranean. This proposal was quickly dismissed by other Atlantic Alliance partners. Speaking in Lisbon in November 1996 Secretary-General Solana announced that the PfP was particular to central and eastern Europe: "It cannot be applied wholesale to the Mediterranean region. In the Mediterranean we can learn from PfP, but we have to find and apply specific solutions."[33] A Partnership for the Mediterranean would be potentially destabilizing if implemented. According to the principle of variable geometry, Israel and NATO could agree to a comprehensive package of measures that Arab dialogue countries would not be able to match. This would create tensions between NATO, Israel, and the other dialogue countries. Moreover, other Arab states not part of the dialogue could be concerned at the prospect of their Arab neighbors strengthening military ties with NATO.[34]

Secretary-General Solana had still left open the possibility of adopting a PfP-inspired model for the Mediterranean in the future. In spring 1998 Portuguese officials were continuing to lobby for the eventual implementation of this model although they were aware that the circumstances then were not favorable. In the meantime, they were attempting to establish closer links between the Mediterranean dialogue countries and the successor of the NACC, the Euro-Atlantic Partnership Council (EAPC). Founded in 1997, the aim of the EAPC was to provide an institutional framework that would help raise the level of political and military cooperation between NATO and countries participating in PfP programs. Within the EAPC, NATO's partners would have an increased say in decision-making. Portuguese officials suggested that the Mediterranean dialogue countries establish links with the EAPC Science Committee. However, many NATO members, preferring to keep PfP activities separate from the Mediterranean dialogue, were reluctant to approve of closer ties between the EAPC and the Mediterranean Initiative.

Greece and Turkey were not opposed to the NATO-Mediterranean dialogue. However, it appeared to come low down in their lists of priorities. Both countries rather focused on their many bilateral problems. Greece was also apprehensive about the security situation in the Balkans. However, Greek leaders had earlier expressed an interest in the CSCM, and the government in Athens traditionally enjoyed close relations with the Arab world. Turkish officials were distracted by developments in the Transcaucasus and Central Asia, particularly with regard to the competition between various states, including Turkey, over the choice of routes for the future transportation of Caspian oil and gas. The authorities in

Ankara were also concerned with the regional dimension of the Kurdish question and the roles of Iraq, Iran, and Syria in particular. With regard to NATO, Turkish officials were giving more attention to the issue of NATO enlargement and were attempting to secure the admission of Bulgaria and Romania in order to enhance the south-east European dimension of the Atlantic Alliance. Nevertheless, Turkey has developed very close military ties with Israel. In January 1998 Turkey, the United States and Israel participated in a search-and-rescue exercise in the Mediterranean, much to the anger of many Arab governments, including Egypt. Interestingly, in 1995 Turkish Defense Minister Vefa Tanir had publicly expressed his support for the Italian proposal for a Partnership for the Mediterranean.[35]

Obviously, the support of the United States was crucial for the launching and continued progress of the NATO-Mediterranean dialogue. The initial reluctance of the United States to support any cooperative initiatives in the Mediterranean that might endanger the Middle East peace process and threaten the continued presence of the US Sixth Fleet had largely disappeared by 1995. However, the Clinton administration was still determined to ensure that the NATO-Mediterranean dialogue not interfere in the workings of the Middle East peace process. Thus, for example, on account of American pressure, the Mediterranean dialogue countries were not permitted to participate in NATO environmental projects as this may have diverted attention from the environmental aspects of the peace process. The Clinton administration was also eager to begin the process of NATO enlargement. Dialogue with the Mediterranean was all well and good for the United States provided it did not distract NATO states from strengthening ties with central and eastern Europe.

It is generally agreed that of the NATO allies, Canada, the Scandinavian countries, and Germany were less enthusiastic about the Mediterranean Initiative. Britain was anxious about the financial costs of the Mediterranean dialogue but was prepared to follow the US policy line. The somewhat anomalous position of France within NATO is well known. It has been suggested that France has not looked warmly at NATO's Mediterranean Initiative and that French officials were rather more willing to lend backing to European initiatives in the area, and in particular, to the EU-sponsored Barcelona Process. It would not be surprising if France was suspicious of NATO's interest in the western Mediterranean given that this has been an area of special traditional concern to France. Certainly in 1996 there was friction between the United

States and France regarding who should preside over a reorganized NATO command structure for the Mediterranean. French officials failed to secure the appointment of a European commanding officer. The French eagerness to pursue its own separate policies in the Mediterranean and in the Middle East was clearly apparent when in October 1996, as part of his tour of the region, President Jacques Chirac became the first foreign head of state to address the Palestine Legislative Council. Given the long-standing French presence in the Mediterranean, it is important for NATO and for the dialogue countries that France assume a greater interest in the Mediterranean Initiative.

THE PERSPECTIVES OF THE MEDITERRANEAN DIALOGUE COUNTRIES

In spite of progress made in NATO's Mediterranean Initiative, the Arab dialogue countries remain suspicious of NATO and the intentions of individual Atlantic Alliance members. They complain that they do not know the aims and objectives of the dialogue and its ultimate goal. Although NATO officials tend to downplay the need to identify definite goals and objectives other than the concern to improve stability and security in the Mediterranean, it would seem that for the Arab dialogue countries this is not enough. They question whether NATO is merely seeking to protect itself from possible risks and threats. And why, ask the Arab dialogue countries, is the dialogue not more "transparent"? What exactly do NATO discussions with Israel entail? The Arab dialogue countries suspect that NATO has no intention of making them genuine partners. They still fear that NATO may attempt to interfere in their internal affairs. The restructuring of NATO force structures, the establishment of EUROFOR and EUROMARFOR—although, as will be discussed in the next chapter, these are technically forces answerable to WEU—and the discussions in NATO circles about the importance of crisis management fuel Arab suspicions of NATO objectives. The Arab dialogue countries also question to what extent NATO speaks with one voice. Are countries like Italy, Portugal, and Spain really concerned about the Mediterranean, or are they merely seeking to exploit Mediterranean security issues in order to bolster their own positions in NATO and in the EU? Given that the Arab dialogue countries have raised these awkward questions, it is perhaps surprising that NATO's Mediterranean Initiative has progressed as far as it has.

Furthermore, there are also serious problems between the Arab dialogue countries and NATO concerning human rights and democracy. Apparently, this was a source of contention between various participants in the NATO-sponsored conference in Rome in November 1997, which was convened to discuss a draft report prepared by the US RAND Corporation on the NATO-Mediterranean dialogue. Some Arab officials complain of double standards. In their opinion, Western governments have no right to lecture their Arab counterparts on human rights issues when these same Western governments have granted asylum to "Islamic terrorists." Arab officials admit that there are shortcomings in the Arab world with regard to the establishment of democracies, but they point out that democratization is a long and arduous process that took centuries to achieve in the West. Therefore, Arab officials expect Western governments to show more tolerance and understanding with regard to what they argue is a steady and gradual movement toward further democratization in many Arab countries, especially in North Africa. These are large issues that cannot be discussed in any detail here. There has been some progress toward the establishment of democracies in some Arab countries, as discussed in detail earlier. Much more political and economic reform is required, though. Differences in culture and in political and social norms also have to be considered. However, these differences should not be used as an excuse to forestall the democratization process in the Arab world.

As already indicated, more confidence and trust between states north and south of the Mediterranean will be gained if Arab governments and their publics become convinced that NATO is not seeking to interfere in the internal affairs of Arab states. Here, recent and more distant historical memories, political, social, and economic differences, and the presence of distinct cultures all have an impact on how governments and peoples perceive one another from across the Mediterranean. Some Arab officials may believe that their worst suspicions have been confirmed when they read an article by American commentators in *NATO Review* that poses the question whether the NATO-Mediterranean dialogue is intended either to shore up Arab regimes or bring about a change in the system of government.[36]

Although Arab dialogue countries have publicly complained that there is not enough "meat" in the NATO-Mediterranean dialogue, perhaps they are inwardly satisfied with the cautious, incremental nature of the NATO-Mediterranean Initiative. Given the lack of confidence and mutual trust between states north and south of the Mediterranean, and given the tensions between Israel and the Arab world, how, for example, would Arab officials explain to their publics a decision to participate in military maneuvers with Israeli and NATO armed forces? It is worth noting in passing, that there are also cynics in the West who believe that the Arab dialogue countries are only interested in participating in the NATO-Mediterranean dialogue in order to improve their relations with Europe and thereby possibly secure more financial support from the EU.

Secretary-General Solana has unsuccessfully attempted to address some of the complaints voiced by the Arab dialogue partners. In a speech in January 1997 he declared that NATO was aiming to dispel some "false ideas that exist in the South about NATO's new roles and institutions. We certainly do not see the Mediterranean as the source of a new threat."[37] Speaking in Rome in November 1997 the secretary-general commented that NATO was not solely interested in the Mediterranean out of concern for political instability. That would be "too narrow and negative." Instead, NATO allies considered the area of significance to the rest of Europe for reasons of trade, investment, maritime transport, natural resources, environmental interdependence, and the problems of human migration inter alia. Solana added that the question of WMD was more a problem for states within the area and less a north-south issue.[38] But the Arab dialogue countries do not seem to have been swayed by these and other speeches.

Egypt regards itself as one of the leading states in the Arab world. Perhaps partly because of a perceived need to demonstrate their credentials, the Egyptian authorities have been especially critical of various aspects of the NATO-Mediterranean dialogue. Mindful of the earlier collapse of the negotiations conducted by ACRS, NATO officials have been careful to ensure that Egyptian-Israeli differences over the issue of the proliferation of WMD should not sabotage the Mediterranean Initiative. The Egyptians are keen to cooperate with NATO in the scientific field, specifically with regard to tackling desertification. Coordinating work with NATO in the area of counterterrorism also attracts Egypt.[39] The Egyptian authorities continue to have difficulties in rooting out radical Islamic terrorists who are threatening to destroy the country's valuable tourist industry. Egyptian officials are also requesting the assistance of NATO in helping to demine large stretches of territory around El Alamein. Pointing out that NATO has considerable expertise in this area and that the Atlantic Alliance is involved in demining activities in Bosnia, it seems that Egyptian officials are making the demining issue a virtual test case for NATO. The question of demining does appear to be of paramount concern for Egyptian officials. They have unsuccessfully sought to secure support from WEU on this issue. Unfortunately for Egypt, and perhaps for the dialogue, NATO officials regard the demining issue as merely of economic rather than crucial humanitarian significance. With the removal of mines, more housing construction could be undertaken. Assisting the Egyptian economy in this manner does not appear to be high on the list of NATO's priorities.

Like Egypt, Morocco has participated in the IFOR and SFOR operations in Bosnia. In March 1996 the prime minister (and also foreign minister) of Morocco, Abdellatif Filali, paid a much-publicized visit to NATO Headquarters.[40] But, as at the start of the NATO-Mediterranean dialogue, the Moroccans remain suspicious of NATO's intentions. They have also been quite outspoken in their desire for Algeria, Lebanon, Libya, and Syria to be included in the dialogue.

The Tunisians share the skepticism of the Moroccans and are also keen that more Arab states be added to the Mediterranean Initiative. They have called for dialogue based on real cooperation rather than simply conducting dialogue for the sake of dialogue. However, as in the case of Morocco, Tunisia has close political and military ties with several NATO member states. In practice, the Tunisian authorities, more fearful of internal threats to their security posed by radical Islamic groups, may be hoping that closer ties with NATO, and particularly with the EU, will help to boost the legitimacy of their regime.

The poorest and most socially and economically backward of the dialogue countries, Mauritania, appears to appreciate its involvement in the Mediterranean Initiative. Still apprehensive about their relations with their giant Moroccan neighbor to their north, perhaps the Mauritanian authorities believe that participating in the NATO-Mediterranean dialogue together with Morocco will considerably ease their security concerns. Mauritanian officials are interested specifically in benefiting from NATO know-how in the fields of water resources and environmental protection.[41] Mauritania would not object if fellow AMU members Algeria and Libya are added to the dialogue.

Jordan is another Arab dialogue country that has expressed little public criticism of NATO's Mediterranean Initiative. The Jordanians have traditionally had close relations with the West, the only aberration being Amman's failure to condemn the Iraqi invasion of Kuwait. Jordan has been an active member of IFOR and SFOR and is a key player in the Middle East peace process. Jordanian officials have expressed a particular interest in benefiting from NATO's expertise in the fields of countering drug smuggling, antiterrorism, and the prevention of man-made disasters. The Jordanian interest in possibly expanding the dialogue to incorporate Gulf countries has been discussed.

The Israeli authorities appear to have no serious problems with the NATO-Mediterranean dialogue. The Mediterranean Initiative is a useful political exercise for the Israelis, as it associates their state with five Arab countries that normalized relations with Israel. Interestingly, Israel does not seem to be playing an active, vocal role in encouraging NATO to deepen and widen the scope of the Mediterranean Initiative. There may be a number of reasons to explain Israeli behavior. Because of the very close military cooperation between Israel and the United States, there is arguably little need for the Israeli armed forces to participate in cooperative activities with NATO in the military field. Arab dialogue countries would probably not react favorably to NATO developing extensive relations with Israel in the information, science, and military fields. Although the principle of variable geometry may be applied in theory in the NATO-Mediterranean dialogue, it appears that both NATO officials and Israeli representatives are aware of the Arab backlash that could ensue if NATO and Israel are perceived to be cozying up to one another. Arab complaints of the lack of transparency concerning the NATO-Israeli dialogue have been previously noted. Nevertheless, Israeli officials are interested in cooperating with NATO in such areas as civil emergency planning and counterterrorism.

CONCLUSION

There is an impression that the NATO-Mediterranean dialogue has steadily progressed since its inception in February 1995, and to a great extent this is true. In spite of the introduction of an element of multilateralization, NATO's Mediterranean Initiative basically remains six separate dialogues between the Atlantic Alliance and the dialogue countries. An increasing number of cooperative activities are open to all dialogue countries, although many of these activities have to be self-funded. In interviews with NATO representatives and officials of the Mediterranean dialogue countries the author has been drawn to conclude that the dialogue is in danger of lapsing into a condition of semi- if not total inertia. In their "discussions," NATO officials often initially wait to listen to the views and observations of the dialogue country. In turn, the representatives of the dialogue country first wait to hear from NATO officials. A dialogue might be effective only if both parties are willing to talk and if at least one of the parties is prepared to take the lead and initiate discussions. If both parties no longer have anything to say to one another, then a dialogue would most probably naturally expire. But this does not appear to be the case with the NATO-Mediterranean dialogue. Instead, the parties concerned seem to be unsure about how to proceed in their discussions. It is to be hoped that this caution, if not reticence, will in time tend to fade as the parties become more involved in a wider range of cooperative activities. Further confidence-building is clearly required.

On perhaps a more positive note, however, at a meeting in Valencia in February 1999, organized by the Spanish Ministry of Defense and the RAND Corporation, it was noted that Egypt, Jordan, and Israel were pressing for the dialogue to be more "region-specific." There was a call for the dialogue to take up an agenda that would focus more on security issues specific to the Mediterranean. Secretary-General Solana in his remarks at the Valencia Conference referred to the absence of a region-specific framework for cooperation.[42]

The author does not share the views of those who contend that the NATO-Mediterranean dialogue, like the WEU-Mediterranean dialogue, is merely a "public relations" exercise and a "confidence-building process," but is not a CBM. Critics argue that NATO's Mediterranean Initiative is not a CBM because it is not contingent on any form of verification and control.[43] Here, the more narrow, "European" definition of CBMs is being employed. One might argue rather that the NATO-Mediterranean dialogue is an initial CBM itself, which has set in motion a confidence-building process that could result in the emergence of further, more advanced CBMs. Arguably, other CBMs are already taking shape within the Mediterranean Initiative. This line of thinking makes use of a more "comprehensive" definition of CBMs that also takes into account the importance of discussions, the exchange of information, and transparency. Therefore, the various cooperative activities that NATO and the Mediterranean dialogue countries have initiated, including even cooperation in the military field, may also be considered as CBMs in their own right. In the confidence-building process that started with the launching of NATO's Mediterranean Initiative, one might argue that more developed CBMs are being gradually introduced.

The NATO-Mediterranean dialogue may hence be regarded as a useful exercise in preventive diplomacy. Clearly, NATO does have something to offer the Mediterranean dialogue countries. It is significant to note that the Mediterranean Initiative, at the time of writing, has not become another victim of problems in the Middle East peace process. However, the NATO-Mediterranean dialogue should be put in perspective. There are other ongoing initiatives and dialogues concerning the Mediterranean, and the most important of these by far is the EU-sponsored Barcelona Process. The next chapter provides a brief overview of the most recent developments in these other cooperative initiatives and examines how the NATO-Mediterranean dialogue might complement these other undertakings.

NOTES

[1]Much of the information and detail in this chapter (not formally cited) stems from interviews conducted with NATO officials, national representatives of Atlantic Alliance delegations at NATO Headquarters in Brussels, and diplomatic representatives from the embassies in Brussels of the NATO dialogue countries.

[2]Press Release (95) 12—February 8, 1995—Statement by the NATO Spokesman on NATO's Mediterranean Initiative.

[3]Press Communiqué M-NAC-1 (95) 48, Ministerial Meeting of the NAC in Noordwijk Aan Zee, The Netherlands, May 30, 1995, paragraph 11.

[4]Andrew J. Pierre and William B. Quandt, *The Algerian Crisis: Policy Options for the West* (Washington, D.C.: The Brookings Institution, The Carnegie Endowment for International Peace, 1996), 45.

[5]M. A. Boumendif, "Aide à la jeune démocratie ou defiance au futur etat islamique—La Récente Initiative de l'OTAN montre une ambivalence dangereuse vis-à-vis de l'Algerie," *La Tribune,* February 12, 1995.

[6]See, for example, the keynote address by Jordanian Crown Prince El Hassan bin Talal to the opening session of the Mediterranean Charter Convention, Madrid, January 10, 1997.

[7]Press Release (95) 12—February 8, 1995.

[8]Jette Nordam, "The Mediterranean Dialogue: Dispelling Misconceptions and Building Confidence," *NATO Review* 45, 4 (July–August 1997): 27.

[9]NAA, Political Committee, Sub-Committee on the Southern Region, *Draft Interim Report,* AM 106 PC/SR (95) 1 by Mr. Rodrigo de Rato (Spain), Rapporteur (Brussels: International Secretariat, May 1995), 7.

[10]*Turkish Daily News,* February 27, 1995; February 21, 1995.

[11]NAA, de Rato Report, May 1995, 6.

[12]NAA, Civilian Affairs Committee, Sub-Committee on the Mediterranean Basin, *Draft Interim Report—Cooperation for Security in the Mediterranean,* AM 83 CC/MB (96) 1 by Mr. Pedro Moya (Spain), Rapporteur (Brussels: International Secretariat, May 1996), 2.

[13]Press Communiqué M-NAC-1 (95) 48, paragraph 11.

[14] Gregory L. Schulte, "Responding to Proliferation—NATO's Role," *NATO Review* 43, 4 (July 1995): 17.

[15] NAA, Civilian Affairs Committee, Sub-Committee on the Mediterranean Basin, *Report—Frameworks for Cooperation in the Mediterranean,* AM 259 CC/MB (95) 7 by Mr. Pedro Moya (Spain), Rapporteur (Brussels: International Secretariat, October 1995), 13.

[16] "Mediterranean Security: New Issues and Challenges." Keynote address by the deputy secretary-general of NATO at the RAND Seminar, October 16, 1995. Note, the NATO Gopher website does not give other details of this speech.

[17] NAA, Moya Report, May 1996, 5.

[18] Press Communiqué M-NAC-2 (95) 118—Ministerial Meeting of the NAC, Brussels, December 5, 1995, paragraph 12.

[19] Final Communiqué of the Ministerial Meeting of the NAC, Brussels, December 10, 1996, paragraph 12.

[20] Press Communiqué M-NAC-1 (97) 65—May 29, 1997—Final Communiqué, Ministerial Meeting of the NAC in Sintra, Portugal, paragraph 6.

[21] Press Communiqué M-1 (97) 81—July 8,1997—Madrid Declaration on Euro-Atlantic Security and Co-operation, issued by the Heads of State and Government Participating in the Meeting of the NAC in Madrid, July 8, 1997, paragraph 13.

[22] Nordam, "The Mediterranean Dialogue," 28.

[23] NAA, Mediterranean Special Group, *Draft General Report—NATO's Role in the Mediterranean,* AP 245 GSM (97) 9 by Mr. Pedro Moya (Spain), General Rapporteur (Brussels: International Secretariat, August 1997), 8, 9, 12.

[24] Speech by NATO Secretary-General Solana, at the Centre Militare di Studi Strategici/RAND International Conference on the Future of NATO's Mediterranean Initiative, Rome, November 10, 1997.

[25] Alberto Bin, "Strengthening Cooperation in the Mediterranean: NATO's Contribution," *NATO Review* 46, 4 (winter 1998): 25–26.

[26] NATO Press Release (98) 44—April 20, 1998—First Course for Mediterranean Partners at the NATO Defence College.

[27] Bin, "Strengthening Cooperation in the Mediterranean," 26–27.

[28] Speech by NATO Secretary-General Solana at the Centre Militare di Studi Strategici/RAND International Conference, November 10, 1997. Solana again stressed that the NATO-Mediterranean dialogue was based on "variable geometry" when addressing a conference on NATO's Mediterranean Initiative in Valencia in February 1999. See secretary-general's Remarks at the Conference on the Mediterranean Dialogue and the New NATO, Valencia, February 25, 1999.

[29] Nordam, "The Mediterranean Dialogue," 27.

30*Foreign Broadcast Information Service—Daily Report—Western Europe (FBIS-WEU-97–339)*, December 5, 1997, Carlos Segovia, "Spanish Ministry on Possible NATO-Maghreb Operations," *El Mundo,* December 5, 1997.

31For a discussion of the views of General Corcione, see the paper by Paolo Pansa Cedrone, "Outline of a West Mediterranean View," in *Report on the Sixth International Antalya Conference on Security and Cooperation,* organized by the Atlantic Council of Turkey, Antalya, November 2–5, 1995, 63.

32Ambassador Antonio Monteiro, "NATO and the Mediterranean Security Dialogue (a PfP-Inspired Model)" (paper presented at the Mediterranean Dialogue Seminar organized by the NAA Mediterranean Special Group, Lisbon, December 5–6, 1996).

33"NATO and the Development of the European Defence and Security Identity." Speech by NATO Secretary-General Solana at the IEEI Conference, Lisbon, November 25, 1996.

34Gréta Gunnarsdottir, "NATO's Mediterranean Dialogue: Current Status—Future Prospects" (paper presented at the Mediterranean Dialogue Seminar organized by the NAA Mediterranean Special Group, Lisbon, December 5–6, 1996).

35Presentation by Turkish Defense Minister Vefa Tanir, in *Report on the Sixth International Antalya Conference on Security and Cooperation,* organized by the Atlantic Council of Turkey, Antalya, November 2–5, 1995, 7.

36Ronald D. Asmus, Stephen Larrabee, and Ian O. Lesser, "Mediterranean Security: New Challenges, New Tasks," *NATO Review* 44, 3 (May 1996): 31.

37Speech by NATO Secretary-General Solana at the Royal Institute of International Relations, Brussels, January 14, 1997.

38Speech by NATO Secretary-General Solana at the Centre Militare di Studi Strategici/RAND International Conference, November 10, 1997.

39NAA, Moya Report, May 1996, 4.

40Ibid., 5.

41Ibid.

42Secretary-general's remarks at the Conference on the Mediterranean Dialogue and the New NATO, February 25, 1999.

43See, for example, Claire Spencer, "Building Confidence in the Mediterranean," *Mediterranean Politics* 2, 2 (autumn 1997): 36.

CHAPTER 8

The Significance of Other Mediterranean Dialogues and NATO's Mediterranean Initiative

INTRODUCTION

One aim of this chapter is to examine the role of other dialogues and initiatives concerning the Mediterranean in the period since 1995. The focus will be on the WEU and OSCE Mediterranean dialogues and the EU's Barcelona Process. Reference will also be made to the activities of the NAA and the Mediterranean Forum. Obviously, it will not be possible to provide a detailed, in-depth analysis of each of these initiatives. Concentration will be on those aspects of these dialogues that may be of significance for NATO's Mediterranean Initiative. The chapter will also discuss how NATO and the aforementioned other institutions and forums might coordinate their respective Mediterranean dialogues and initiatives.

NATO officials have often emphasized the need for their Mediterranean Initiative to "complement" but not "duplicate" other ongoing Mediterranean dialogues, and they argue that this objective is being accomplished.[1] Duplication would be pointless and would lead to a considerable waste of resources, time, and energy. Complementing other initiatives sounds much more reasonable, but begs the question as to how NATO might complement the activities of other institutions and forums.

Officials at NATO have noted that the NATO-Mediterranean dialogue "complements" the efforts of the EU's Barcelona Process, the OSCE, and WEU "without creating any division of labour."[2] Speaking in Rome in November 1997, NATO Secretary-General Solana stated that although there was some "overlap" of the different Mediterranean initiatives, a "comprehensive strategy" for the Mediterranean clearly required that responsibility be shared by all institutions in a "complementary way."³ Overlap, therefore, was unavoidable and inevitable. In practice, this would entail a measure of duplication. Presumably, it was believed that this would not create difficulties, as each dialogue would remain separate and distinct. A rigid division of labor would have avoided overlap. However, one may assume that NATO officials were opposed to such a division of labor because it would set limits on their freedom of maneuver. NATO policy-makers would not have warmed to the idea of being instructed by an outside party as to what areas NATO's Mediterranean Initiative may or may not apply.

Bearing in mind their close links, the coordination of activities of NATO and the NAA was the least problematic. The NAA has continued with its own Mediterranean dialogue, which was launched at a seminar in Paris in March 1995. At the second NAA Mediterranean Dialogue Seminar, held in Lisbon in December 1996, delegates from Cyprus, Egypt, Israel, Jordan, Mauritania, Morocco, the Palestinian territories, and Tunisia participated. A third Mediterranean Dialogue Seminar, organized in Istanbul in November 1997, and a fourth in Cairo in December 1998 were also well attended. Because of the increasing importance of the Mediterranean, in May 1996 the NAA replaced its Sub-Committee of the Mediterranean Basin with a new Mediterranean Special Group (MSG). The purpose of the MSG was to improve coordination of the NAA's activities in the Mediterranean. The MSG was composed of representatives from the NAA's Civilian Affairs, Political, and Defense and Security Committees.

The NAA has been attempting to coordinate its activities with NATO's Mediterranean Initiative. The MSG held a brainstorming session immediately before the NAA's Mediterranean Dialogue Seminar in Lisbon. In the seminar it was agreed that the priority should be to oversee and accompany NATO's efforts to redefine its role in the Mediterranean. The MSG decided to produce a document that would explain to the Mediterranean dialogue countries of the NAA the roles and functions of NATO.⁴ Earlier in 1996, NAA circles had suggested that the Mediterranean dialogues of the NAA and NATO could be more closely linked by offering permanent observer status in the NAA to those NATO Mediterranean dialogue countries that did not have such status. Jordan, Mauritania, and Tunisia were the countries concerned. According to criteria previously agreed to by NAA officials, one may presume that a case could be argued that each of these states were "emerging as democracies." However, apparently, Tunisia in particular expressed little interest.⁵

NATO officials clearly believe that there is room for closer cooperation with other institutions and forums concerning the Mediterranean. The importance of complementing other initiatives in the area is continually referred to in NATO texts. But, NATO officials have no formal mandate to coordinate their activities involving the Mediterranean with other organizations or bodies. They are aware that NATO's Mediterranean Initiative may only operate within certain bounds. Secretary-General Solana has admitted that the EU is the key player in the Mediterranean given that most of the security challenges in the area stem from deteriorating social and economic conditions. Thus, according to Solana, the EU's Barcelona Process is the central multilateral initiative involving the Mediterranean.[6] Nevertheless, the actual and potential role of NATO in the Mediterranean, and the possibilities for NATO to complement effectively the activities of other institutions and forums, should not be downplayed.

THE WEU-MEDITERRANEAN DIALOGUE AND NATO'S MEDITERRANEAN INITIATIVE

In May 1998 Jordan became a party to the WEU-Mediterranean dialogue. Apparently, Greece, in particular, had lobbied for Jordan's inclusion. The presidency and secretariat of WEU continues to meet regularly with the ambassadors in Brussels of the non-WEU Mediterranean countries. This dialogue is still basically limited to the seven principles previously agreed on, which focus exclusively on military/defense aspects of security. In these discussions WEU officials also talk of the significance for the Mediterranean of WEU's so-called Petersberg tasks—humanitarian and rescue tasks, peacekeeping missions, and tasks of combat forces in crisis management including peace-making. The ambassadors are also briefed on the latest WEU Ministerial Meeting.

However, the dialogue with experts of the non-WEU Mediterranean countries that was launched by the WEU Mediterranean Group in fall 1994 had come to a halt after a final meeting in October 1995. Thus, Israeli experts had only participated in one of these gatherings. Meetings were terminated because little had been achieved. Issues discussed in meetings with experts had already previously been handled in the ambassadors' dialogue. The WEU Mediterranean Group continues to function.

The WEU's Institute for Security Studies based in Paris has been more successful. The Institute organizes an annual Mediterranean seminar that is attended by representatives from the non-WEU Mediterranean countries. The seminar held in June 1997 was titled "National and International Approaches to Peacekeeping among the Euro-Mediterranean Countries." Officials and military officers from Algeria, Egypt, Israel, Mauritania, Morocco, and Tunisia participated. At the seminar the North Africans approved of Anglo-French and EU/WEU proposals for assisting the OAU in its peacekeeping missions. The representatives from North Africa, though, stressed that the UN should remain the most important body, and they also underlined the need for noninterference in the internal affairs or the work of regional bodies.[7]

The WEU Assembly—a body of deputies from WEU member countries that makes recommendations and supervises the work of the WEU Ministerial Council—has failed to build ties with parliamentarians from the Mediterranean. It is an associate participant to the work of the IPU, which organized interparliamentary conferences on security and cooperation in the Mediterranean at Malaga in 1992 and Valletta in 1995. In spite of the recommendations of a WEU Report published in 1995, the WEU Assembly has been unable to grant guest or observer status to parliamentarians of non-WEU Mediterranean countries.[8] Apparently, the subject is still under review. A major problem, though, is the lack of available resources at the disposal of the WEU Assembly.

Overall, little progress has been achieved in the WEU-Mediterranean dialogue. Unlike its NATO counterpart, the WEU-Mediterranean dialogue has failed to develop practical programs of cooperation with Mediterranean countries. WEU is not able to benefit from the experience and make use of the resources of NATO PfP-type programs. In the second half of 1997, under the presidency of Germany, WEU did organize a visit for representatives of non-WEU Mediterranean countries to WEU's satellite center at Torrejon in Spain—but this appeared to be a one-off event. Egypt's requests for WEU assistance in demining stretches of the Egyptian desert have been turned down. As in the case of NATO, there are divisions in WEU between the Club Med grouping—in WEU's case, France, Italy, Portugal, and Spain—and the northern Europeans who are more in favor of consolidating ties with WEU's Associate Partners in central and eastern Europe. In a rather pessimistic assessment, a WEU Report issued in November 1996 noted that "given the WEU has not become involved in the political aspects, the limited dialogue on which it has embarked with these (Mediterranean) countries has not produced any tangible results." The Report added that the WEU's involvement in the Mediterranean was welcomed by Morocco, but was a disappointment for Egypt and Turkey (an Associate Member).[9]

A major problem for WEU in general, is the continued uncertainty with regard to the future of the WEU-EU relationship. The EU Amsterdam Summit in mid-1997 revealed once again that a CFSP for Europe would not be realized in the short or even medium term. It was decided at Amsterdam to develop further the security and defense dimension of the EU. The Petersberg tasks were incorporated into the Treaty on EU, and it was noted that WEU could be made use of for the preparation and implementation of these tasks. There were also references to building up WEU in stages as the defense component of the EU.[10] However, unanimity was still required for decisions having military or defense implications. And in the WEU-EU relationship it was clear that WEU remained subordinate. There are many references in WEU and EU official texts that speak of occasions when the EU might choose to "avail itself of WEU." The EU may or may not choose to make use of WEU with regard to future policies that may have security and defense implications. The announcement made by British and French ministers in St. Malo in late 1998 that the EU should in future have credible military forces, cast into further doubt the future of WEU and its relationship with the EU.

The WEU-Mediterranean dialogue does not appear to have benefited from the EU's Barcelona Process. It seems that WEU has largely assumed the role of a passive bystander as the Barcelona Process evolves. WEU Ministerial Council Communiqués have declared that "the WEU Mediterranean Group will contribute its expertise within the framework of its general mandate in response to requests from EU."[11] Again, there is no obligation for the EU to make such requests. Indeed, WEU failed to secure a seat at the inaugural Barcelona Conference in November 1995. It has been suggested that this was because if WEU had participated the Arab League would have pressed to be accorded the same status. Some observers feel that the political and security chapter of the Barcelona Process is being developed at the expense of the WEU-Mediterranean dialogue.[12] Will the further evolution of the Barcelona Process make the WEU-Mediterranean dialogue redundant? Apparently, when EU delegations visit non-WEU Mediterranean countries, they do not include WEU officials in their ranks as participants or observers. As a reaction to this turn of events, in 1998 officials of WEU openly declared their readiness to offer their expertise to the EU to help develop the political and security chapter of the Barcelona Process.[13]

In general, WEU and NATO have been increasingly coordinating their activities since the first formal meeting of high-level WEU officials and the NAC in Brussels in May 1992. Naval vessels of WEU and NATO operated together in the Adriatic during the crisis in Bosnia. At Maastricht in 1991 there had been much talk of the need for WEU and NATO to build up their relations on the basis of "transparency and complementarity." The two bodies were encouraged to coordinate in areas of common interest.[14] Much time and energy has been consumed attempting to work out how NATO assets might be employed for future possible WEU-led CJTFs. This type of CJTF could involve the implementation of a Petersberg task in territory traditionally regarded as out-of-area.

Both WEU and NATO officials have also referred to the need to coordinate their Mediterranean dialogues. In May 1996, in Birmingham, WEU Ministers welcomed the fact that WEU and NATO were holding joint meetings to discuss the Mediterranean.[15] One year later in Paris WEU Ministers agreed to keep the WEU-Mediterranean dialogue under review and to take into account the role of other initiatives such as the Barcelona Process and NATO's Mediterranean dialogue.[16] Similar wording was used by WEU Ministers in Rhodes in May 1998.[17] The Communiqué of the NAC meeting in Berlin in June 1996 spoke of identifying additional areas of "focused NATO-WEU cooperation" including joint meetings of their respective dialogues.[18] Contrary to the text of the meeting of the WEU Ministerial Council in Birmingham, it seemed that according to the NAC as of June 1996, no joint meeting between NATO and WEU on their respective Mediterranean dialogues had taken place.

There has been little mention of the WEU-Mediterranean dialogue in NATO texts since June 1996. This may be because of the lack of progress in the WEU-Mediterranean dialogue. Coordination between the dialogues of NATO and WEU would be complicated by the fact that the WEU-Mediterranean dialogue includes Algeria. Apparently, though, when the Algerian ambassador in Brussels attends meetings of the WEU-Mediterranean dialogue, he is interested to learn news of NATO's Mediterranean Initiative from WEU officials.

Concerning the Mediterranean dialogues of WEU and NATO it is also important to note the roles of EUROFOR and EUROMARFOR. The announcement that these Euroforces were to be established was issued at the WEU Council of Ministers meeting in Lisbon in May 1995. They were referred to as "forces answerable to WEU." These forces were to consist of units provided by France, Italy, Portugal, and Spain, although in theory other WEU states could join and allocate troops and equipment to the Euroforces if this was agreed to by the four founding members.[19] These were to be nonstanding forces. It was envisioned that EUROFOR would be a rapid-reaction force of ten to fifteen thousand troops, and

EUROMARFOR would be a joint naval force with aeronaval and amphibious elements. Both forces could be deployed to implement the so-called Petersberg tasks. In 1996 the Euroforces were declared operational. EUROFOR was allocated a headquarters in Florence with a multinational staff working on planning and eventual command and control. The use of either of the Euroforces would need the consent of all four contributing states. The Euroforces could be deployed separately, or combined. They could be called upon by WEU, or by other bodies such as NATO, the OSCE, or the UN.[20] In November 1997 WEU ministers meeting at Erfurt stated that Italy and Spain had decided to establish a joint Amphibious Force that would be answerable to WEU. This force could also be used by NATO.[21]

In spite of the close connections between WEU and the Euroforces, it seems that basic coordination is lacking. On May 13, 1997, WEU officials decided that the Euroforces could assume responsibility for those parts of the WEU-Mediterranean dialogue that concentrated on crisis management and early warning. Occasional meetings are thus held at WEU Headquarters in Brussels between representatives of the Euroforces and diplomats from the non-WEU Mediterranean countries. However, it seems that in practice these meetings are conducted outside the framework of the WEU-Mediterranean dialogue as WEU officials are not well informed about conclusions reached at these meetings. Is there a danger that the WEU-Mediterranean dialogue will become even more insignificant as a consequence of the discussions conducted by officials of the Euroforces?

Initially, NATO policy-makers reacted somewhat coldly to the announcement of the future establishment of the Euroforces. On May 30, 1995, the NAC noted at Noordwijk: "We look forward with interest to a high-level briefing on this initiative and to the expeditious definition of the relationship of these forces with WEU and NATO."[22] It seems that NATO officials had not been properly consulted. Officials at NATO were anxious to ensure that possible future operations carried out by the Euroforces would not prejudice the participation of these units in missions to defend NATO territory. NATO wanted first call on the Euroforces in the event that territory of the Atlantic Alliance was threatened or attacked. The Lisbon Declaration had stated rather equivocally that the Euroforces, in addition to being forces answerable to WEU, could also be employed in the framework of NATO.[23] Eventually, NATO officials were assured that if required they could have first call on units allocated to EUROFOR and EUROMARFOR.

The Arab world reacted harshly to the announcement of the establishment of the Euroforces and to news of the opening of the EUROFOR headquarters in Florence. It seems that WEU officials had not adequately consulted beforehand with representatives of the non-WEU Mediterranean countries.[24] The Palestinian newspaper *al-Quds al-Arabi* speculated that the "prime task" of EUROFOR was to intervene in Libya. The Libyan news agency JANA said that the existence of EUROFOR was a "flagrant violation of the sovereignty of southern Mediterranean states." The Libyan leader, Gadhafi, declared that the opening of the EUROFOR Headquarters was a "declaration of war against the Arab states south of the Mediterranean," was tantamount to a "terrorist attack," and was a "threat to peace in the Mediterranean region." The then Tunisian Defense Minister Habib Ben Yahia stated that the creation of EUROFOR was "incompatible with the traditions of dialogue and talks which Tunisia is seeking to establish between the two sides of the Mediterranean." Referring generally to the Euroforces, Egyptian President Mubarak commented: "the issue needs explanations," and added, "I fear that it opens the way to interference in other states' internal affairs."[25]

Such statements must have a negative impact on NATO's Mediterranean Initiative, especially when bearing in mind that Arab officials tend to link the Euroforces with NATO and not WEU even though the forces are technically answerable to WEU. There is a general Arab concern that NATO might use the Euroforces as instruments to interfere in the internal affairs of Arab states. This fear is understandable given that the Euroforces are equipped to implement the Petersberg tasks. Arab reaction to the now infamous article published in December 1997 in the Spanish newspaper *El Mundo* has been discussed. With regard to the Euroforces, there is little NATO officials may do directly to assuage Arab anxieties. They can encourage the Euroforces to continue to attempt to improve their public image. The Euroforces do have their own dialogue with non-WEU Mediterranean countries. Arab experts have been invited to observe WEU military exercises—apparently, Morocco took up the invitation—and visit the EUROFOR headquarters in Florence. In a declaration issued as early as May 1997, the defense and foreign ministers of the member states of the Euroforces announced that the Euroforces were prepared to cooperate in future with other Mediterranean countries to implement any of the Petersberg tasks in the Mediterranean. Cooperation was thus envisioned in such areas as human and natural disasters, maritime policing, the protection of sea lanes and merchant shipping, minesweeping.[26] The role of the Euroforces could be discussed in detail in the NATO-Mediterranean dialogue.

THE OSCE-MEDITERRANEAN DIALOGUE AND NATO's MEDITERRANEAN INITIATIVE

There have been no dramatic developments in the OSCE-Mediterranean dialogue since the landmark Budapest Summit in December 1994. However, Jordan was admitted as a new member to the dialogue in May 1998. In November 1995 the OSCE secretary-general had suggested that Jordan and Mauritania—both parties to the NATO-Mediterranean dialogue by that time—be included in the OSCE-Mediterranean dialogue.[27] As envisioned at Budapest, there have been ministerial-level meetings between representatives of the original five (now six) nonparticipating Mediterranean states and the OSCE troika and secretary-general. The first of these was held in July 1995 in Vienna. There have also been regular, more informal meetings of the OSCE Contact Group and representatives of the Mediterranean states as outlined at Budapest. In this format, for example, participants have discussed how the OSCE's work on a Comprehensive Security Model for the Twenty First Century would not be complete without the full integration of the Mediterranean dimension. In April 1997 a briefing was given to Mediterranean officials by the OSCE chairman-in-office. As well as describing current OSCE activities, the briefing also covered the topic: "Military Aspects of Security; How to Promote CSBMs."[28] Annual OSCE Seminars on Mediterranean issues have been organized. The first seminar, held in Cairo in September 1995, focused on the OSCE experience in confidence-building. The second, convened in Tel Aviv in June 1996, discussed the role of the OSCE as a platform for dialogue and the fostering of norms of behavior. In September 1997 a third seminar, again held in Cairo, examined the implications for the Mediterranean of the OSCE's Comprehensive Security Model. A fourth seminar, convened in Valletta in October 1998, discussed the human dimension of security.

Again, in line with the decisions agreed upon at Budapest in December 1994, the representatives of the nonparticipating Mediterranean states have been invited to attend other meetings in which issues pertaining to the Mediterranean are discussed. Thus, in June 1995 the five contributed to the third meeting of the OSCE's Economic Forum, which met in Prague. "Regional" economic cooperation in the fields of trade, investment, and infrastructures was analyzed. The five were invited to attend meetings of the Economic Forum on a regular basis. They also participated in OSCE Ministerial Meetings. For example, at the Ministerial Meeting in Budapest in December 1995, the Egyptian delegate suggested that the OSCE embark on a program of cultural CBMs in order to prevent cultural misunderstandings that could hinder attempts to enhance security in the Mediterranean area.[29]

At the December 1995 Plenary Meeting of the Permanent Council in Budapest, it was decided that the five Mediterranean states would henceforth be referred to as "Mediterranean Partners for Cooperation." This sounded better than the much less flattering previous designation—"non-participating Mediterranean states." The change in title was most probably on account of pressure from the five Mediterranean states. They had been seeking to enhance their relationship with the OSCE. A special meeting of the OSCE Council devoted to Mediterranean issues had been held in November 1995 with the participation of the North Africans and Israel. In practice, though, nothing substantive had been achieved with the change of title. Indeed, the Budapest Plenary had stated: "These changes in name do not alter the specific relationships between these states and the OSCE set out in previous OSCE decisions."[30] There was no upgrading of the dialogue, although the issue was apparently raised again just before the OSCE Lisbon Summit in December 1996. In June 1998, though, the Mediterranean Partners for Cooperation were invited to send observers in future to elections monitored by the OSCE. They were also encouraged to send visitors to OSCE missions.[31] However, it still seems that the priorities of the OSCE lie elsewhere, in central and eastern Europe, in the Balkans, and even in Central Asia. The OSCE tends to put the Mediterranean on the same level of importance as Japan and South Korea.[32]

The purpose of the OSCE-Mediterranean dialogue is far from clear. Is the OSCE seeking to develop a "regional" initiative for the Mediterranean within the framework of the OSCE? This could assume the form of a Mediterranean regional roundtable, for example.[33] Or, rather, is the OSCE attempting to present itself—and its predecessor, the CSCE—as some sort of model that the southern Mediterranean states can emulate? The emphasis here would be on encouraging the improvement of interstate relations in the southern Mediterranean and providing more stability and security within these states. In this case the Mediterranean Partners for Cooperation would remain outside of the OSCE framework. At the OSCE Ministerial Council Meeting held in Copenhagen in December 1997, when agreeing to the guidelines for a new OSCE Charter for European Security, ministers declared that they would consider cooperation with all partners "in order to promote the norms and values shared by the OSCE participating states. They will also encourage partners to draw on OSCE expertise."[34] This leads one back to the question of how relevant are particular principles and norms, developed in Europe, for the southern and eastern Mediterranean. One might contend that the OSCE experience is of value for the Mediterranean area where the Arab states and Israel could also make use of certain institutions modeled on OSCE practice.

If the OSCE was to serve as some sort of inspirational model for the southern and eastern Mediterranean states, OSCE officials could place emphasis on the relevance for the Mediterranean of the original Declaration of Principles of the then CSCE. Military CBMs concerned with preventing a surprise attack are more relevant for interstate relations in the southern and eastern Mediterranean than for a north-south framework. In the current circumstances it is exceedingly unlikely that any state in the southern Mediterranean would launch a surprise attack against its neighbors to the north, or vice versa. The experience of the CSCE in fostering cooperation between European states in the Cold War period by focusing on three baskets of cooperation was envisioned by the stillborn CSCM and is being applied by the EU's Barcelona Process in its trans-Mediterranean cooperation initiative. As discussed in the second chapter of this study, conflict prevention mechanisms along the lines of the OSCE's Conflict Prevention Center and the Forum for Security Cooperation could be employed within a north-south framework in the Mediterranean, and/or could be applied along a south-south/Arab-Israeli dimension. These mechanisms, for instance, could help promote transparency by encouraging the exchange of information, could provide a forum for negotiation in order to tackle problems, and could handle the issue of unusual military activities. But, as noted, if the troubled Gulf area is excluded, actual conflict is more likely to erupt between states in the southern or eastern Mediterranean. As indicated earlier, the OSCE's Code of Conduct on Politico-Military Aspects of Security is not likely to appeal to the Arab states, in particular, because of the Code's stress on the need for the civilian control of military, paramilitary, and other security forces.

Since the beginning of the Middle East peace process, certain "regional" initiatives for the "Middle East"—however this may be defined—have been floated. Article Four of the Israeli-Jordanian Peace Treaty of 1994 committed both parties to setting up a Conference on Security and Cooperation in the Middle East (CSCME).[35] A CSCME could be modeled on the CSCE experience and would presumably be exclusively concerned with the states of the Middle East—unlike the earlier envisioned

CSCM. There has also been talk of a possible Organization for Cooperation in the Middle East (OCME), which could also be modeled on the CSCE/OSCE experience, and which again would presumably only incorporate states in the Middle East. The possible establishment of an OCME was raised by British Foreign Secretary Malcolm Rifkind on his visit to the United Arab Emirates in November 1996. According to Rifkind, an OCME would gradually evolve to include all states in the Middle East, including "pariah" states such as Iraq and Iran. Diplomats admitted that it would be difficult to establish cooperation on issues like human rights, frontier disputes, and national minorities.[36]

The OSCE in its Mediterranean dialogue has not attempted to make use of the CSCE/OSCE experience in Europe and apply it to the Mediterranean. Rather ACRS and then the EU's Barcelona Process have attempted to employ elements of the CSCE/OSCE approach in Europe to the Mediterranean—in a north-south and south-south interstate dimension, with possible implications for the intrastate relations in the south. In these circumstances, it appears that there is little scope for the further development of the OSCE-Mediterranean dialogue. This perhaps explains why NATO officials have not paid much attention to this dialogue. Coordination would anyway be difficult because of the inclusion of Algeria and the exclusion of Mauritania from the OSCE-Mediterranean dialogue. If, in future, the OSCE's Mediterranean Partners for Cooperation were somehow incorporated more fully within the OSCE framework, NATO officials would most probably be interested in coordinating activities. However, it seems unlikely that the OSCE-Mediterranean dialogue will evolve along these lines. NATO officials are thus much more interested in the progress of the Barcelona Process. In particular, they are closely following the military/political chapter of the Barcelona Process that is attempting to draw on CSCE/OSCE experiences. Given the commencement of NATO's cooperative activities in the military field with its Mediterranean dialogue countries, it would seem that there is an increasing need for NATO and the EU to coordinate their policies in the Mediterranean.

THE BARCELONA PROCESS AND NATO's MEDITERRANEAN INITIATIVE

Following the decisions taken at the meeting of the European Council in Essen in December 1994, the Barcelona Conference was eventually held on November 27–28, 1995. Foreign ministers of the fifteen member states of the EU and of the twelve invited Mediterranean noncommunity countries participated. Most significantly, in contrast to their boycott of the multilateral talks in the Middle East peace process, Lebanon and Syria had agreed to attend in spite of the presence of Israel. Mauritania and the United States attended as observers. Despite lobbying from the AMU, Britain and France refused to allow Libya to participate. The Libyans then withdrew their request to join the conference and accused the organizers of "high treason."[37] The Barcelona Conference spoke of a Euro-Mediterranean partnership. The Barcelona Declaration issued at the end of the conference declared that the objective was to turn the Mediterranean basin into "an area of dialogue, exchange and cooperation guaranteeing peace, stability and prosperity." The need to respect various principles such as human rights, democracy, respect for the territorial integrity and sovereignty of other states, and the peaceful settlement of disputes was emphasized. Along the lines of the CSCE and the aborted CSCM, the participants agreed to concentrate their future cooperation in three broad areas or "chapters" (i.e., "baskets")—namely, politics and security; economics and finance; and social, cultural, and human relations.[38]

The Barcelona Conference set in motion the so-called Barcelona Process. A follow-up work program was outlined. Future meetings were to be held at ministerial and senior-officials level. This would lead to intergovernmental discussions on issues such as water resources, energy policy, industry, tourism, and environment. More informal gatherings of NGOs representing civil society were encouraged. The Barcelona Process was to be monitored by the Euro-Mediterranean Committee for the Barcelona Process that consisted of officials of the current EU troika and representatives of the twelve Mediterranean non-Community countries. The Committee would meet every three months and report to the foreign ministers. The European Parliament also initiated contacts with deputies of Mediterranean-partner assemblies and thereby launched an interparliamentary dialogue with the support of the IPU.

The second chapter of the Barcelona Declaration, which focused on economic and financial partnership, referred to the aim of creating a free trade zone in the Euro-Mediterranean area by the year 2010. In order to further this goal, the EU would continue to conclude new association agreements with its Mediterranean partners.[39] Immediately prior to the Barcelona Conference the EU had negotiated agreements with Morocco and Tunisia. Similar agreements with Israel and Jordan were later concluded. Here, the EU was seeking to consolidate its bilateral ties with

Mediterranean partners and thereby also strengthen and develop multilateral cooperation with regard to free trade. Economic and financial cooperation between states north and south of the Mediterranean would also be boosted by the decision of the European Council at Cannes in June 1995 to provide southern and eastern Mediterranean states ecu 4.68 billion in support for the period 1996–2000—note, the European Commission had originally hoped to extend ecu 5.5 billion. The Barcelona Declaration also underlined the need for "regional" cooperation between the Mediterranean partners themselves that would help to create a free-trade area for the Mediterranean as a whole. The attempts by the AMU to foster regional cooperation and the later AFTA initiative have been noted. Enhanced cooperation in the Euro-Mediterranean area in the fields of trade and investment was also recommended in the Declaration. The EU could attempt to make use of its coordination of REDWG, one of the five working groups set up in the multilateral framework of the Middle East peace process. Agreements concluded at the MENA economic conferences could also have had beneficial consequences for the second chapter of the Barcelona Process. But problems in the Middle East led to the suspension of the work of REDWG and prevented the holding of a MENA economic conference in 1998.

However, the Barcelona Declaration only referred to a free trade area in the Euro-Mediterranean area with regard to industrial goods and services. Concerning agricultural products, preferential and reciprocal access was envisioned. The Mediterranean partners were encouraged to promote free trade among themselves. The non-EU twelve had pushed for a genuine free trade area throughout the Euro-Mediterranean area that would have enabled their agricultural products to enter the EU without restrictions.[40] Under the arrangement that was agreed upon, the Mediterranean partners will be disadvantaged. They will have to lower tariff barriers that provide protection for local manufacturing companies and thus be exposed to fierce competition from Europe.[41] There appears to be a general hope that in the interim twelve-year period investment will flow southward and help to create jobs and boost private enterprise. This would require economic reforms and political and social stability in the southern and eastern Mediterranean states. But, as discussed previously, the prospects for further political and economic development in North Africa are uncertain at best. It has been suggested that a compensation fund could be created to assist those who suffer most as reforms are implemented.[42] Some studies suggest that up to 40 percent of existing jobs may be lost in the southern and eastern Mediterranean. Local

The Significance of Other Mediterranean Dialogues 211

companies, unable to attract foreign investment and failing to modernize, may be forced to close.[43]

It does not seem likely that in the foreseeable future a Euro-Mediterranean Development Bank will be established as a result of progress in the Barcelona Process. The United States was frustrated by one delay after another in its efforts to make use of the MENA economic conferences to set up a Middle East Development Bank in Cairo. The Americans have also encountered opposition from the EU that has preferred to lobby for the creation of a so-called Middle East and North Africa Financial Intermediary Organization (MENAFIO). According to EU officials, MENAFIO would focus on mobilizing private-sector financing for investment projects and infrastructural development—assuming that investors are prepared to part with their money. These officials have argued that there are already a number of development banks in the "region" willing to lend financial support.[44]

The third chapter of the Euro-Mediterranean partnership initiative is titled: "Partnership in Social, Cultural and Human Affairs: Developing Human Resources, Promoting Understanding between Cultures and Exchanges between Civil Societies." In this chapter the Barcelona Declaration stressed the importance of cooperation between states north and south of the Mediterranean in such areas as educational and cultural programs, activities covering the mass media, and exchange schemes involving the leaders of political and civil society, the cultural and religious worlds, universities, trade unions, the business community, and so forth. Here, use could be made of cooperative programs the EU had established earlier with the Mediterranean, such as MED-CAMPUS, MED-MEDIA. As part of a very broad remit, the third chapter was also supposed to cover other issues such as terrorism, drug-trafficking, crime and corruption, migration, and the need to promote democratic institutions and strengthen the rule of law.

The Euro-Med Civic Forum has come to play an active role in establishing links between civil societies across the Mediterranean. The first meeting of the Civic Forum, held immediately after the Barcelona Conference in Barcelona, which was attended by over one thousand participants, established eleven parallel working forums.[45] The Euro-Med Civic Forum has continued to meet. The topics that were discussed at its session in Naples in December 1997 included: the Mediterranean area and globalization; value and culture as resources; relations and communications strategies; and economic and social interrelationships. Also, as a part of the Barcelona Process, but separate from the Euro-Med Civic

Forum, a Workshop on the Dialogue between Cultures and Civilizations was held in Stockholm in April 1998.[46] In order to develop the parliamentary dimension of the Barcelona Process, a Euro-Med Parliamentary Forum held its first meeting in Brussels in October 1998. By early 1999 a Euro-Med Youth Program had also been launched.

It will not be easy to bridge the cultural divide across the Mediterranean. But increased exchange of information and expertise could help to overcome certain misperceptions and prejudices that individuals and groups may hold vis-à-vis their counterparts north or south of the Mediterranean. Bearing in mind the political, social, and economic problems of states in the southern and eastern Mediterranean, the development of a more confident and mature civil society willing to persuade governments to introduce or implement certain reforms could be encouraged. However, there is a danger that general progress in the third chapter may be stalled by specific points of complaint. Migration—perhaps better suited to the second chapter—is, potentially, one such issue. There has been vague talk of reducing migratory pressure by forming training programs and job creation schemes.[47] However, the Mediterranean partners prefer the freer movement of peoples. They are not keen on EU attempts to impose tighter immigration controls.[48] There are also problems, once again, over differing interpretations concerning human rights and democratic norms. And the distinction between terrorist groups and those fighting for liberation and independence has also proven to be a bone of contention.

The Mediterranean Forum could help the consolidation of the third chapter of the Barcelona Process even though the EU appears to have picked up many of the educational, cultural, and social themes that the Forum had intended to develop. The Mediterranean Forum has continued to hold annual ministerial meetings. Malta was admitted as a new member at the meeting in Sainte-Maxime in southern France in April 1995. Other states such as Albania, Croatia, Cyprus, Israel, Libya, Mauritania, Slovenia, and also Jordan and Russia (the latter two, geographically, not Mediterranean states) have pressed to join. One of the strengths of the Mediterranean Forum is the informal nature of its proceedings. A report of the session at Sainte-Maxime noted that the Mediterranean Forum was "[a]n exercise avoiding overlapping with other already existing institutions."[49]

At Sainte-Maxime, French Foreign Minister Alain Juppe declared that the Barcelona Process and the Mediterranean Forum were two independent but complementary exercises, although the latter was less ambitious and more modest.⁵⁰ Interestingly, the draft RAND Report on NATO's Mediterranean Initiative released in September 1997 suggested that because of the informal nature of the Mediterranean Forum, issues such as terrorism and the proliferation of WMD could be discussed more freely in the Forum than within the framework of the Barcelona Process. In future, therefore, it is possible that the EU's Barcelona Process may attempt to build up closer working relations with the Mediterranean Forum.

The first chapter of the Euro-Mediterranean partnership initiative is titled: "Political and Security Partnership: Establishing a Common Area of Peace and Stability." In this chapter, the Barcelona Declaration referred to the problems of the proliferation of WMD, terrorism, organized crime, and drug-trafficking. The Declaration stipulated that one aim was to secure "a mutually and effectively verifiable Middle East zone free of weapons of mass destruction, nuclear, chemical and biological, and their delivery systems." Parties were encouraged to refrain from developing their military capacity beyond "legitimate defence requirements" and to reduce their amounts of troops and weaponry to the lowest possible limits. The parties also undertook to.

> consider any confidence and security-building measures that could be taken between the parties with a view to the creation of an "area of peace and stability in the Mediterranean," including the long term possibility of establishing a Euro-Mediterranean pact to that end.

Clearly, this was a highly ambitious menu of CBMs, CSBMs, arms limitations measures, and weapons free zones.

The second meeting of foreign ministers of the parties involved in the Barcelona Process, which assembled in Malta in April 1997, no longer referred to a Euro-Mediterranean pact but rather spoke of the need to draw up a Mediterranean Charter for Peace and Stability. The Malta meeting failed to agree to guidelines and principles for this Charter. An Arab position paper indicated that because of current difficulties in the Middle East peace process the Charter could not be adopted. Arab delegates could see little sense in agreeing to a Charter that sought to prevent future conflicts but avoided tackling ongoing disputes.⁵¹ Arab representatives also wanted the Charter to focus on issues such as the proliferation of WMD, and socioeconomic problems including the issues of migration and debt.⁵² The security concerns of Arab governments were clearly in evidence here. The Arab parties wanted the Mediterranean

Charter to address the question of Israeli nuclear weapons. They were also eager to consolidate their governments at home by securing political support and economic backing from Europe. The Mediterranean partners were opposed to the idea of modeling the Mediterranean Charter on the Pact on Stability in Europe that had been adopted by European states in March 1995. Arab states, and also Israel and Turkey, were uneasy about some of the provisions of the Pact that focused on minority issues and state borders.

The third meeting of foreign ministers involved in the Barcelona Process, which gathered in Stuttgart in mid-April 1999, was finally able to consider "Guidelines for Elaborating a Euro-Mediterranean Charter." These were submitted as an informal working document. References were made to the importance of developing so-called partnership-building measures, good neighborly relations, and regional cooperation and preventive diplomacy. However, apparently, much work remains to be done. A Group of Senior Officials will continue to work on preparing the Charter before the next ministerial conference. The intention is for ministers to approve the Charter "as soon as political circumstances allow."[53]

The second Euro-Mediterranean ministerial conference at Malta was unable to endorse two other key documents relevant to the first chapter of the Barcelona Process. One of these was a Plan of Action that covered six areas—the strengthening of democracy, preventive diplomacy, security and CBMs, disarmament, terrorism, and organized crime. In an effort to obtain EU funds, Egypt had also raised the issue once again of mine clearance.[54] Reference to the need to strengthen democracy probably aroused some concern among the Mediterranean partners. It was not clear what this would entail. Could some form of mechanism be adopted that could enable parties to intervene in the internal affairs of other states?[55] In those circumstances would not Arab governments perceive that the Europeans were attempting to impose upon them certain norms and values? The Plan of Action's reference to the need to strengthen democracy throughout the Euro-Mediterranean area certainly challenged the notion that the status quo should be respected and maintained in certain states. No mention was made of a Plan of Action in the text of the Conclusions of the third Euro-Mediterranean ministerial conference at Stuttgart in April 1999.

The foreign ministers at Malta also failed to agree to draw up an inventory of CSBMs. The intention of establishing a conflict-prevention center was put on hold. However, the setting up of a mechanism for cooperation in the event of natural and human disasters was agreed upon in principle. At Stuttgart, progress toward establishing a Euro-Mediterranean system of disaster prevention, mitigation, and management was noted as an example of a partnership-building measure. Various CBMs in the areas of information exchange and increased transparency were already in place before the Malta conference. EU-sponsored diplomatic seminars have been organized. Contact points among the twenty-seven partner states have been erected for exchange of information on political and security matters. Also, a network of international affairs and strategic studies institutes in the region, known as EuroMeSCo, has been established. An informal EuroMeSCo–Senior Officials seminar titled the "Euro-Mediterranean Security Dialogue" was held in Bonn in March 1999.

What are the prospects for the further development of the political-military chapter of the Barcelona Process? It has been suggested that the parties could implement measures that ACRS had been in the process of introducing—such as the installation of an effective communication network that could help prevent future conflicts in the region. More developed CBMs may then follow. With regard to interstate relations in the south in particular, these may include the formation of demilitarized zones and areas in which military activities are limited. States in the south may agree to adopt nonoffensive military postures. They may also promise not to deploy new weapons systems that could destabilize regional security.[56] In the more immediate term, a Euro-Mediterranean maritime coast guard could be formed to deal with narcotics trafficking, the transport of illegal migrants, and maritime pollution.[57]

There are a number of problems relating to the Barcelona Process as a whole. It is extremely difficult to secure a consensus among the twenty-seven parties. In spite of the size of membership, certain key states are not included. Without the inclusion of Libya, Iran, and Iraq, it will not be possible to create a Middle East zone free of WMD.[58] It is important to note, though, that a Libyan delegation was invited to attend the third Euro-Mediterranean ministerial conference at Stuttgart as a guest of the presidency. It was declared that Libya could become a full member of the Barcelona Process as soon as UN sanctions were lifted and after Libya had accepted the whole of the Barcelona *acquis*.[59]

The Barcelona Process has also been criticized because of its overemphasis on north-south issues at the expense of tackling problems related to interstate and intrastate relations in the south.[60] However, the Barcelona Declaration did stress the need to develop regional cooperation between the southern partners. And, as already indicated, with regard to developments within a state, Mediterranean partners are not enthusiastic about being told how to reform their systems of government.

It appears that there will be no dramatic progress in the political-security chapter of the Barcelona Process while serious problems remain in the Middle East. Arab states will not agree to a comprehensive package of CBMs in the Euro-Mediterranean area until a final Arab-Israeli peace settlement has been concluded according to the formula of "land for peace." The Barcelona Declaration had envisioned that confidence-building would occur in incremental stages.[61] Arab governments are opposed to implementing more advanced CBMs out of a concern that this would freeze the current status quo, which is perceived to be in Israel's favor. The major differences of opinion between the Arab states and Israel concerning the proliferation of WMD have also stalled progress in the first chapter. It is open to question how much progress can be realized in the second and third chapters of the Barcelona Process if there is little or no movement in the political-security fields.

The failure of the EU to agree to a CFSP is another general problem for the Barcelona Process. It is not always possible to secure a consensus among the fifteen EU members on certain issues. The EU has attempted to coordinate policy in the Middle East with the onset of the peace process. The EU has offered extensive financial aid to the Palestinians. Much Arab capital is invested in EU member states.[62] The groundbreaking visit of French President Jacques Chirac to the area in October 1996 led to the appointment of an EU envoy to the Middle East. However, the role and responsibility of this envoy is not clear. There is a limit to what the EU is able to accomplish in the Middle East without the backing of the United States, which is not a full party to the Barcelona Process. One may assume that the Americans are in favor of the Barcelona Process provided that it does not complicate and pose problems for the Middle East peace process. In the Barcelona Declaration and the Conclusions of the Malta and Stuttgart meetings the Euro-Mediterranean partners have openly declared that the Barcelona Process is not intended to replace "the other activities and initiatives undertaken in the interest of peace, stability and development in the region." It seems that this so-called self-denying ordinance was intended to assure the Americans.[63] In practice, the Israelis and many Arabs would not be prepared to embark on new policies in the Middle East without consulting beforehand with the US administration.

How might NATO's Mediterranean Initiative be coordinated with the EU's Barcelona Process? There are already apparently many informal links between officials of NATO and the EU based in Brussels. Speaking in Valencia in February 1999 Secretary-General Solana declared that NATO and the EU could achieve more in the Mediterranean if there was "more coherence between their policies."[64] Activities organized as a part of the NATO-Mediterranean dialogue appear to complement the work of the second and third chapters of the Barcelona Process. There would seem to be more overlap concerning the political-military chapter. However, duplication is not possible because of the absence of the United States in the Barcelona Process. Writing in 1996, Tanner suggested that a possible Euro-Mediterranean pact dealing with abstract norms-building could co-exist with NATO PfP-type activities in the Mediterranean. Tanner also proposed links with the WEU-Mediterranean dialogue and possibly with the OSCE's Conflict Prevention Center in Vienna.[65] From a vantage point three years later, the WEU-Mediterranean dialogue has not developed, but it is still possible to envision future coordination in the Mediterranean between NATO, with its interest in promoting concrete forms of cooperation in the information, science, and military fields, and the Barcelona Process, with its concern to develop other forms of CBMs, CSBMs, and conflict prevention measures.

CONCLUSION

There is thus much potential for coordination between the NATO-Mediterranean dialogue and the EU's Barcelona Process. With regard to other initiatives and dialogues concerning the Mediterranean, NATO and the NAA are likely to continue to closely cooperate. In future, NATO officials may be more interested in learning about the discussions conducted between officers of the Euroforces and representatives of non-WEU Mediterranean countries. However, policy-makers at NATO will probably pay less attention to the WEU-Mediterranean dialogue, which has scarcely developed since its inception. It is not clear how NATO may complement the activities of the Mediterranean Forum and the OSCE. The Mediterranean Forum may rather work increasingly in tandem with the EU's Barcelona Process. Officials in the OSCE need to work out in what direction they would prefer to see the OSCE-Mediterranean dialogue evolve. It is possible, for instance, that the OSCE may seek to develop ties with its Mediterranean partners in the social and cultural fields and thus complement the NATO-Mediterranean dialogue. The impact on the Barcelona Process of CSCE/OSCE norms and principles and certain OSCE institutions has been discussed.

Of course, discussion on how NATO might coordinate its activities in the Mediterranean with other dialogues and initiatives is dependent on how NATO's Mediterranean Initiative evolves in the next years. The future of the Middle East peace process is also of crucial importance for all of the dialogues and initiatives herein examined.

NOTES

[1] See, for example, Press Communiqué M-NAC-2 (95) 118—Final Communiqué of the Ministerial Meeting of the NAC, Brussels, December 5, 1995, paragraph 12.

[2] Final Communiqué of the Ministerial Meeting of the NAC, Brussels, December 10, 1996, paragraph 12.

[3] Speech by NATO Secretary-General Javier Solana at the Centri Militare di Studi Strategici/RAND International Conference on the future of NATO's Mediterranean Initiative, Rome, November 10, 1997.

[4] NAA, Mediterranean Special Group, *Chairman's Conclusions and 1997 Work Programme, 4 December 1996, Lisbon, Portugal,* AN 319 GSM (96) 1 (Brussels: International Secretariat, December 1996), 1–3.

[5] NAA, Standing Committee, *Restructuring NAA Mediterranean Activities,* AN 33 SC (96) 9 by Pedro Moya (Spain), Rapporteur of the Sub-Committee on the Mediterranean Basin and Anders Sjaastad (Norway), Chairman of the Defence and Security Committee (Brussels: International Secretariat, March 1996), 3.

[6] Speech by NATO Secretary-General Javier Solana at the Centri Militare di Studi Strategici/RAND International Conference, November 10, 1997.

[7] *Institute for Security Studies, WEU Newsletter,* 21, October 1997, 2.

[8] The Report in question is, *Assembly of Western European Union, Doc.1485, 6 November 1995,* "Parliamentary Co-operation in the Mediterranean." Report submitted on behalf of the Committee for Parliamentary and Public Relations by Mr. Kitsonis, Rapporteur, 17.

[9] *Assembly of Western European Union, Proceedings, 42nd Session, December 1996, I Assembly Documents* (Paris: WEU), Doc.1543, November 4, 1996, "Security in the Mediterranean Region." Report submitted on behalf of the Political Committee by Mr. de Lipkowski, Rapporteur, 157–58.

[10] See, WEU, *Ministerial Council, WEU Council of Ministers, Declaration*

of WEU on the Role of WEU and Its Relations with the EU and with the Atlantic Alliance, Amsterdam, 22 July 1997.

[11]See, for example, WEU, *Ministerial Council, WEU Council of Ministers, Paris Declaration, Paris, 13 May 1997*, paragraph 41; and WEU, *Ministerial Council, WEU Council of Ministers, Erfurt Declaration, Erfurt, 18 November 1997*, paragraph 48.

[12]Fred Tanner, "The Mediterranean Pact: A Framework for Soft Security Cooperation," *Perceptions* 1, 4 (December–February 1996/97): 65.

[13]Carlos Echeverria, *Cooperation in Peacekeeping among the Euro-Mediterranean Armed Forces* (Paris: Chaillot Papers 35, Institute for Security Studies, WEU, 1999), 32.

[14]"Declaration of the Member States of Western European Union which are also Members of the European Union on the Role of the Western European Union and Its Relations with the European Union and the Atlantic Alliance," in *WEU Related Texts adopted at EC Summit, Maastricht—10 December 1991* (London: WEU Press and Information Section, December 10, 1991).

[15]WEU, *Ministerial Council, WEU Council of Ministers, Birmingham Declaration, Birmingham, 7 May 1996*, paragraph 18.

[16]WEU, *Ministerial Council, WEU Council of Ministers, Paris Declaration, Paris, 13 May 1997*, paragraph 41.

[17]WEU, *Ministerial Council, WEU Council of Ministers, Rhodes Declaration, Rhodes, 15 May 1998, Doc. 1612*, paragraph 49.

[18]Press Communiqué—Final Communiqué of the Ministerial Meeting of the NAC, Berlin, June 3, 1996, paragraph 20.

[19]WEU, *Council of Ministers, Lisbon Declaration, Lisbon, 15 May 1995, Doc. 1455*, paragraph 5.

[20]For more details of EUROFOR and EUROMARFOR, see *Assembly of Western European Union, Doc. 1468, 12 June 1995*, "European Armed Forces." Report submitted on behalf of the Defence Committee by Mr. de Decker, Rapporteur. See also, "EUROFOR-EUROMARFOR—A Brief Note," by Rafael Estrella, available from the NAA Secretariat, Brussels.

[21]WEU, *Ministerial Council, WEU Council of Ministers, Erfurt Declaration, Erfurt, 18 November 1997*, paragraph 36.

[22]Press Communiqué M-NAC-1 (95) 48—Ministerial Meeting of the NAC in Noordwijk Aan Zee, The Netherlands, May 30,1995, paragraph 8.

[23]WEU, de Decker Report, June 1995.

[24]Claire Spencer, "Building Confidence in the Mediterranean," *Mediterranean Politics* 2, 2 (autumn 1997): 36.

[25]*Al Ahram,* November 21–27, 1996; and, Karen Dabrowska, "The Mediterranean: Cradle of Armed Conflict," *Dialogue* (June 1997): 3.

[26]Echeverria, *Cooperation in Peacekeeping among the Euro-Mediterranean Armed Forces,* 35.

[27]*OSCE Ref.MC/11/95 30 November 1995—OSCE—The Secretary General, Annual Report 1995 on OSCE Activities,* v. Relations with Non-Participating States, 28.

[28]*OSCE Newsletter,* 3, 10 (October 1996): 2; 4, 4 (April 1997): 6.

[29]*OSCE Ref. MC/22/95 7 December 1995—Statement of Egypt presented to the Fifth Ministerial Meeting of the OSCE, Budapest, 7–8 December 1995.* Delivered by Fatih El Shazly, deputy minister of foreign affairs.

[30]*OSCE Permanent Council, PC. DEC/94 5 December 1995* —49th Plenary Meeting. PC Journal 49, Agenda item 4, Decision 94.

[31]*OSCE Permanent Council, PC.DEC/233 11 June 1998*—172nd Plenary Meeting. PC Journal 172, Agenda item 7, Decision 233.

[32]Dominic Fenech, "The Relevance of European Security Structures to the Mediterranean (and Vice Versa)," *Mediterranean Politics* 2, 1 (summer 1997): 172.

[33]The idea of a Mediterranean regional roundtable was suggested in the OSCE Mediterranean seminar held in Cairo in September 1995. See *OSCE Ref. SEC/288/95 16 October 1995 (Vienna)—OSCE Seminar on the OSCE Experience in the Field of Confidence-Building (Consolidated Summary), Cairo, 26–28 September 1995*—Jure Gaspanic, "Code of Conduct, Confidence-Building in the Mediterranean."

[34]*OSCE Ref. MC (6) DEC/5 19 December 1997—MC (6) Journal 2,* Agenda item 8, Decision 5, Guidelines on an OSCE Document, Charter on European Security, 4, paragraph 5 (j).

[35]Joel Peters, *Pathways to Peace: The Multilateral Arab-Israeli Talks* (London: The Royal Institute of International Affairs, 1996), 73–74.

[36]*Guardian Weekly,* November 10, 1996.

[37]Esther Barbé, "The Barcelona Conference: Launching Pad of a Process," *Mediterranean Politics* 1, 1 (summer 1996): 34.

[38]*Barcelona Declaration Adopted at the Euro-Mediterranean Conference, 28 November 1995* (Barcelona: 1995).

[39]Stephen C. Calleya, "The Euro-Mediterranean Process after Malta: What Prospects?" *Mediterranean Politics* 2, 2 (autumn 1997): 11.

[40]Claire Spencer, "A Tale of Two Cities," *The World Today* 53, 3 (March 1997): 80.

[41]Jon Marks, "High Hopes and Low Motives: The New Euro-Mediterranean Partnership Initiative," *Mediterranean Politics* 1, 1 (summer 1996): 2.

42Calleya, "The Euro-Mediterranean Process after Malta," 10.

43Eberhard Kienle, "Destabilization through Partnership? Euro-Mediterranean Relations after the Barcelona Declaration," *Mediterranean Politics* 3, 2 (autumn 1998): 7–8.

44Marks, "High Hopes and Low Motives," 17; and "Middle East Development Bank: A Folly in the Making," *The Economist,* October 28, 1995.

45Barbé, "The Barcelona Conference," 29; n.8, 41–42.

46For details advertizing the Euro-Med Civic Forum meeting and the Stockholm workshop, see, *Euro-Mediterranean Partnership—Monthly Information Notes,* November 1997 edition.

47Sarah Collinson, *Shore to Shore—The Politics of Migration in Euro-Maghreb Relations* (London, and Washington, D.C.: The Royal Institute of International Affairs and The Brookings Institution, 1996), 79.

48Barbé, "The Barcelona Conference," 36, 38.

49*Mediterranean Forum Summary Report—Meetings of High Officials, Sainte-Maxime, 8–9 April 1995.*

50*Réunion ministerielle du forum Méditerranéen—Allocation d'ouverture de ministre des affaires étrangeres, M. Alain Juppe, Sainte-Maxime,* April 8, 1995.

51Spencer, "Building Confidence in the Mediterranean," 44. For the text of the Conclusions of the Second Euro-Mediterranean Ministerial Conference, Malta, April 15–16, 1997, see the appendix to Calleya, "The Euro-Mediterranean Process after Malta," 15–22.

52Fred Tanner, "The Euro-Med Partnership: Prospects for Arms Limitations and Confidence Building after Malta," *The International Spectator* 32, 2 (April–June 1997): 6.

53*Chairman's Formal Conclusions of the Third Euro-Mediterranean Ministerial Conference, Stuttgart, 15–16 April 1999* (Stuttgart: 1999). See paragraphs 10–12.

54Tanner, "The Euro-Med Partnership," 4–5, 19.

55Roberto Aliboni, "Confidence-Building, Conflict Prevention and Arms Control in the Euro-Mediterranean Partnership," *Perceptions* 2, 4 (December 1997–February 1998): 80.

56Tanner, "The Euro-Med Partnership," 11, 22.

57Calleya, "The Euro-Mediterranean Process after Malta," 6.

58Tanner, "The Euro-Med Partnership," 22–23.

59*Chairman's Formal Conclusions of the Third Euro-Mediterranean Ministerial Conference,* paragraph 37.

60Aliboni, "Confidence-Building, Conflict Prevention and Arms Control," 79.

61Tanner, "The Euro-Med Partnership," 14.

[62] Rosemary Hollis, "Europe and the Middle East: Power by Stealth?" *International Affairs* 73, 1 (January 1997): 21–23.

[63] Spencer, "Building Confidence in the Mediterranean," 39.

[64] Secretary-general's remarks at the Conference on the Mediterranean Dialogue and the New NATO, Valencia, February 25, 1999.

[65] Tanner, "The Mediterranean Pact," 66–67.

CHAPTER 9
Conclusion

The aim of this chapter is not to recapitulate in detail the main points raised in each of the earlier chapters. The intention, rather, is to focus on examining the prospects for NATO's Mediterranean Initiative, bearing in mind the current situation in the Mediterranean. The assumption, here, is that in the foreseeable future there will be no major crisis in the Mediterranean and that the conflict in Kosovo will not result in a larger Balkan war. It is important to bear in mind, however, that an outbreak of hostilities between Turkey and Greece could seriously damage the Mediterranean Initiative. A complete collapse of the Middle East peace process would probably result in the freezing of the NATO-Mediterranean dialogue.

One should not forget, first, that largely due to its proximity to the Persian Gulf and its overlap with the Middle East, the Mediterranean is an area of strategic importance for NATO in the post–Cold War era. In the ongoing restructuring of NATO commands, it was decided at a meeting of NATO defense ministers in Brussels in December 1997 that a new Regional Command (RC) South should replace AFSOUTH. In addition to having two core components, RC South will have four Joint Sub-Regional Commands (JSRCs) based in Greece, Italy, Spain, and Turkey.[1] Differences between Greece and Turkey have largely been resolved concerning responsibility for control of military flights over the Aegean. In spite of vehement French lobbying, a U.S. commander will be in overall command of the new RC South. Unlike France, Spain intends to participate fully in the new military structure as soon as possible. In spite of earlier lobbying, an Allied Command Mediterranean was thus not established. Only two Strategic Commands (SCs)—for Atlantic and for Europe—were agreed upon. The new RC South is one of two RCs for SC Europe.

After the launching of the EU's Barcelona Process, there is much positive talk concerning the prospects for enhanced security and stability in what is referred to as the "Euro-Mediterranean area." Certainly, in the post–Cold War era the peoples and governments of states immediately to the north and south of the Mediterranean are becoming increasingly more interconnected through various economic, social, and political ties. Issues of security—defining security in its broadest sense—in the southern Mediterranean cannot be totally separated from the security concerns of states north of the Mediterranean. Arguably, in effect, the EU's Barcelona Process is attempting to promote regionalism by encouraging cooperation between governments and peoples in the Euro-Mediterranean area across a wide range of issues.

It is far too early, though, to speak of a "Euro-Mediterranean region." A clearly identifiable Mediterranean region has yet to emerge with its own particular agenda and a measure of autonomy. It is even too early to state that a Mediterranean region is in the making in spite of the activities and work of the EU's Barcelona Process. Although trans-Mediterranean links are expanding, the Mediterranean remains a divide between Europe and the largely Arab North Africa and Middle East. This will remain so as long as there are major socioeconomic disparities between states north and south of the Mediterranean. Important differences in the nature of the political systems also hinder the consolidation and expansion of trans-Mediterranean ties. Recent events, memories of the more distant past, and the absence of a shared culture—culture, here, defined as a set of socially created and learned norms, standards, rules or collective mental programming making a meaningful existence possible for members of a community—also have caused strains in relations between governments and peoples north and south of the Mediterranean. These distinctions and divisions will not disappear or markedly change in the short or medium term.

The primary purpose of the EU's Barcelona Process, the NATO-Mediterranean dialogue, and the other ongoing Mediterranean initiatives is not to engage in region building. The main objective, rather, is to improve relations between governments and peoples in the Mediterranean or Euro-Mediterranean area. In effect, the aim is to move from the current situation of a stable, cold peace to one of a durable, warm peace. The focus, here, is on relations between the European north and the largely

Arab and Muslim south. CBMs are seen as necessary. The preconditions needed for a confidence-building process to develop appear to be in place. There are also incentives for governments and peoples to cooperate. Dialogue will play an important role in this process.

It remains to be seen how effective dialogue will be in establishing what has been referred to as "a culture of cooperation and dialogue" in the Mediterranean.[2] A common "culture of cooperation and dialogue" would help to overcome suspicion and mistrust and lead to a better understanding between governments and peoples in the Euro-Mediterranean area. At the same time, however, differences in "national cultures" would still exist within the embracing umbrella of this common "culture of dialogue and cooperation." But these cultural differences would be less likely to lead to official and popular misperceptions of the policies and interests of governments and peoples on either shore of the Mediterranean. Peoples north and south of the Mediterranean would then tend to view each other positively and this would in turn affect how these peoples interrelate.

Of course, it is important to bear in mind that the security and stability of the Euro-Mediterranean area does not only entail relations between governments and peoples north and south of the Mediterranean. For example, this study has also discussed the significance of south-south interstate relations and the relevance of internal security problems in the Arab states and in Israel. Problems in the Balkans and tensions between Greece and Turkey have not been the focus of this study, as NATO's Mediterranean Initiative does not directly address those concerns.

The NATO-Mediterranean dialogue is primarily concerned with addressing the north-south dimension of Mediterranean security. Although it is much narrower in scope and far less ambitious than the EU's Barcelona Process, it seems that NATO's Mediterranean Initiative can and already is performing a useful role in the Mediterranean or Euro-Mediterranean area. In the discussions and programs of cooperative activities that constitute NATO's Mediterranean Initiative, officials of the Atlantic Alliance are establishing and developing ties not only with governmental representatives, but with military officers, parliamentarians, scientists, academics, and other so-called opinion leaders in the south. As in the case of the initiatives of the EU, OSCE, and the Mediterranean Forum, the NATO-Mediterranean dialogue is attempting to tap into what is an emerging civil society in North Africa and the Middle East. Perhaps, in the longer term, as an indirect spin-off, NATO's Mediterranean Initiative might also hope to improve interstate relations in the south—

for example, Arab and Israeli delegates have grown accustomed to attending together NATO briefings. And contacts with the various spokesmen of civil society may go a little way toward encouraging further democratization in the Arab world, although, as previously noted, in the short term the process of democratization may contribute to instability if ruling elites resist political and economic reforms.

Specific problems in NATO's Mediterranean Initiative—which one should remember is still in its infancy—have been noted and discussed in earlier chapters. There are only six non-NATO Mediterranean countries involved in the dialogue. "Rogue states" such as Libya and Syria have been deliberately excluded. In order to avoid encountering the same fate as ACRS, the dialogue has intentionally skirted seriously addressing the issue of the proliferation of WMD in the Euro-Mediterranean area and beyond. The Arab participants in the dialogue are still unsure of the aims and intentions of NATO.

There is a debate about whether NATO officials should attempt to multilateralize the NATO-Mediterranean dialogue. At the time of writing, NATO only holds occasional joint briefings to which all six parties to the dialogue are invited to attend as a group. Analysts at the RAND Corporation, in the official publication of their 1997 draft study, have suggested that in addition to meeting in a "16 + 1" format, the MCG should also gather with the Mediterranean-dialogue countries once a year in a "16 + 6" format. An NAA Report has proposed that the dialogue should be multilateralized when dealing with issues that could not be as effectively handled between NATO officials and representatives of the Mediterranean-dialogue countries at a bilateral level. The proliferation of WMD was singled out as one such issue that could be possibly better tackled at a multilateral level.[3] Actually, there is a mandate for NATO officials to introduce more multilateralization in the NATO-Mediterranean dialogue. However, for the foreseeable future, it is highly unlikely that talks will be held in a "16 + 6" format to discuss political matters. At present, none of the six Mediterranean dialogue countries are in favor of such a format.

It has been suggested that in order for there to be a foundation for genuine multilateralism, four basic dimensions should be present. First, with regard to the "substantive" dimension, there should be recognition among a group of states of the importance of certain goals, values, and norms. Second, there should be "will and commitment" among the states to take measures to uphold these common goals and values. The third dimension concerns the degree to which a collective enterprise is institutionalized in order to protect these norms and principles. Finally, for multilateralism to work there must be an agreed-upon mechanism and procedure for consultation between the states involved.[4]

In his discussion of multilateralism, Ruggie has likewise emphasized the importance of common values and norms when referring to the need for a group of states in their relations with one another to follow "generalized organizing principles." He adds that there should be an expectation of "diffuse reciprocity" among the parties concerned.[5] Thus, these parties would expect reciprocal equal benefits from the arrangement over a period while, in the meantime, foregoing possible short-term gains.

With reference to the Mediterranean-dialogue countries, in spite of the activities of the EU's Barcelona Process, the NATO-Mediterranean dialogue itself, and other initiatives, there is no shared commitment to abide by certain common goals, values, and norms. NATO, through its Mediterranean Initiative, has endeavored tentatively to promote the fourth of these dimensions, but major progress here cannot be expected until there is a shared commitment to uphold certain values and norms. It seems that the six Mediterranean-dialogue countries are still rather more interested in pursuing their own particular short-term interests. At best, NATO could perhaps encourage multilateral cooperation among the six on specific, technical issues such as civil-emergency planning.

Therefore, it would seem that in the present circumstances, there is little prospect of a real multilateralization of the NATO-Mediterranean dialogue. The non-NATO states in the dialogue do not wish to be referred to as a "group" of six or five states. The five Arab participants do not wish to be too closely associated with Israel. The Arab parties themselves have difficulty in forming a consensus on various issues. This is not surprising given the inability of Arab states in general to cooperate with one another. The failure to develop the AMU is here a case in point. The collapse of the ACRS talks and the considerable difficulties the EU's Barcelona Process has confronted indicate that developing multilateral cooperation with regard to security issues in the Euro-Mediterranean area is a far from simple task. The extent of multilateralism within NATO itself must also be questioned when, for example, noting the serious tensions in Greek-Turkish relations. The Atlantic Alliance is not a monolith. Differences of opinion among NATO allies concerning how much importance should be placed on expanding relations with the non-NATO Mediterranean has been discussed earlier.

There has been talk in some NATO circles of applying to the

Mediterranean the experience the Atlantic Alliance has acquired in expanding its ties with central and eastern Europe. Since the end of the Cold War, NATO has built up its relations with ex–Warsaw Pact states by establishing the NACC and its successor the EAPC. PfP programs have been worked out with individual states. Poland, Hungary, and the Czech Republic were even admitted as new members of the Atlantic Alliance in spring 1999. This study has noted that NATO communiqués and NAA reports referred to the need to create NACC-type structures for the south, and that various Western officials have lobbied for the adoption of PfP-type arrangements for non-NATO Mediterranean states. In practice, though, it is not feasible for NATO officials to transfer lock, stock, and barrel to the Mediterranean those bodies and institutions that have helped NATO successfully develop relations with states in central and eastern Europe.

The NACC was a forum for dialogue and consultation and was a vehicle through which NATO allies and ex–Warsaw Pact states could work together on various projects. There was cooperation in various political, economic, military, environmental, and scientific fields. In practice, though, the NACC had been only a consultative body. It had no operational power, no official budget, no central management, and no real secretariat. The NACC was founded in 1991 at a time when the governments and peoples of the central and east European states, in general, were eager to associate themselves closely with NATO. These states were rapidly democratizing and introducing economic reforms. Their militaries were under the control of the civilian authorities. The ex–Warsaw Pact states were also willing to work with one another and with NATO allies in the NACC, partly out of a fear of possible renewed Russian expansionism. From the establishment of the NACC the emphasis was on building a partnership between NATO and the non-NATO states. The EAPC is in effect an upgraded and institutionalized version of the NACC. For example, it provides an institutionalized framework for cooperation in enhanced PfP activities in the military field. As in the case of its predecessor, the EAPC is a well-functioning multilateral body.

Clearly, the circumstances of the North African and the Middle Eastern states bear no relation to those of central and eastern Europe. A PfP-inspired model for the Mediterranean is also extremely unlikely for the foreseeable future for the reasons outlined previously. PfP-type arrangements would be based on individual packages, thereby obviating the need to cultivate multilateral ties, and the loaded term "partnership" could be dropped for non-NATO Mediterranean states. However, the likely extensive package Israel and NATO would agree on would disturb Arab-dialogue countries. PfP-type arrangements would also be difficult to apply to the Mediterranean because of their likely references—as in the case of the central and eastern Europeans—to the need for the respect of human rights, democratic principles, and the importance of civilian control over the military. Moreover, in many cases the PfP packages negotiated with individual central and eastern European states were regarded as essential prerequisites before these states could be admitted to NATO as full members. In contrast, membership in NATO is obviously not a realistic proposition for states in North Africa and in the Middle East.

How, then, will NATO's Mediterranean Initiative possibly evolve in the foreseeable future, assuming that no major crises erupt in the Euro-Mediterranean area? Realistically, one can only look ahead to the immediate future. In the medium to longer term, the situation in the Mediterranean may be radically altered perhaps as a consequence of succession problems or government change in key states such as Algeria, Egypt, Jordan, Morocco, and Syria.

It is extremely unlikely that NATO member states will reach a consensus and agree to invite other states to join NATO's Mediterranean Initiative. States such as Morocco and Tunisia have been lobbying for the participation of Algeria in the dialogue. The circumstances behind the election of Bouteflika as president in April 1999, however, will not have encouraged NATO member states to push for the inclusion of Algeria in the Mediterranean Initiative. New members to the dialogue would bring along new problems. Given the extent to which NATO has developed relations with the current six Mediterranean dialogue countries, it would be likely that any future newly admitted states would need time to build up the same level of working ties with the Atlantic Alliance. This would make it even more difficult to apply the principle of nondiscrimination where what is offered to one state by NATO in its Mediterranean Initiative should also be offered to other dialogue countries. It would appear that NATO officials believe that six dialogue countries are a manageable number to deal with. These officials are fully aware of the difficulties that the EU's Barcelona Process is encountering partly because of the large number of states involved in that initiative. Moreover, NATO policymakers already have enough problems on their hands in attempting to work more effectively with the current six dialogue countries.

How will discussions within the NATO-Mediterranean dialogue proceed? It has been noted that it seems that the six dialogue countries have tended to assume a passive role. It appears that they have generally preferred to respond to proposals introduced by NATO officials. Israeli delegates, sensitive to the suspicions and concerns of the Arab participants, are not likely to assume a much more active role in these discussions. Given that Egypt regards itself as one of the leaders of the so-called Arab world, it would seem that the onus is on Egyptian officials to take up a more energetic line. The issue of the proliferation of WMD will probably continue to be omitted from these discussions. NATO officials do not have a mandate to discuss this issue extensively within the framework of the NATO-Mediterranean dialogue. These officials are perhaps hoping that the further strengthening of the side of the dialogue concerned with fostering cooperative activities will as a natural spin-off lead to more topics discussed with the dialogue countries and thereby contribute to continued confidence-building between all the parties involved.

It is probable that the cooperative activities between NATO and the Mediterranean-dialogue countries will increasingly focus more on military-related issues. After all, this is a field in which NATO has unparalleled expertise. Speaking in Valencia in February 1999, Secretary-General Solana pressed for NATO and the Mediterranean-dialogue countries to work to develop their military cooperation further. He proposed that additional CBMs in the military domain be explored. These could include the participation of the dialogue countries in peace support and other military-related activities.[6] But, as long as the Arab publics remain suspicious of NATO, Arab governments will remain extremely reluctant to participate in joint military exercises with the Atlantic Alliance. Top-ranking Arab military officers in certain countries may also oppose possible indirect pressure from NATO to allow the civilian authorities to establish more control over the armed forces. However, enhanced cooperation in the military field should help to ease Arab concerns over references made by NATO officials to the need for the Atlantic Alliance to develop crisis-management instruments. It must be emphasized, though, that more money is required for NATO to finance its various programs of cooperative activities with the Mediterranean-dialogue countries.

As outlined in the previous chapter, the NATO-Mediterranean dialogue is likely to be increasingly more coordinated with the EU's Barcelona Process as both initiatives continue to gradually evolve.

The RAND Corporation in its work on NATO's Mediterranean Initiative has come up with a string of recommendations with regard to how NATO officials might further develop their ties with the Mediterranean-dialogue countries. In their 1997 report—published in 1998—the RAND Corporation suggested that as a part of NATO's information activities, the magazine *NATO Review* also be published in Arabic. The report also recommended, as a new feature of the Mediterranean Initiative, the setting up of a joint working group on CBMs that would consist of representatives from the dialogue countries and from a select group of NATO member states. This new body would examine how to develop further CBMs in the Mediterranean.[7] Presumably, such a joint working group would need to coordinate its activities in line with progress made within the Barcelona Process.

In a later report to be released in 1999, members of the RAND Corporation apparently have further advanced their proposals. They are pressing for NATO to establish a crisis-prevention and confidence-building network for the Mediterranean. It appears, though, that much more progress is required in the Middle East peace process before such a network will be realized.

NATO's Mediterranean Initiative will most probably continue to develop incrementally. There will be no major dramatic breakthroughs in relations between states north and south of the Mediterranean solely on account of the work of the NATO-Mediterranean dialogue. However, this dialogue is a useful CBM and an important tool of preventive diplomacy in the Euro-Mediterranean area. It will continue to proceed quietly. In contrast to the NATO enlargement issue, NATO's Mediterranean Initiative will not attract much publicity. With reference to the Mediterranean specifically, it will not command the attention that scholars and commentators have bestowed on the Barcelona Process. Nevertheless, the actual and potential significance of the NATO-Mediterranean dialogue should not be underestimated. NATO member states themselves should lend the dialogue more political and financial support. NATO's Mediterranean Initiative should not be perceived, particularly by the Mediterranean dialogue countries themselves, as merely a hobbyhorse of a select number of NATO allies—the Club Med grouping. The continued interest of the United States is crucial. Likewise, France should lend the NATO-Mediterranean dialogue more active backing. It is also important that Germany be perceived as committed to the stability and security of the Mediterranean as well as to that of central and eastern Europe. However, Mediterranean-dialogue countries also must shoulder some of the responsibility. The NATO-Mediterranean dialogue will not effectively evolve without the backing of governments and other increasingly influential groups in Arab societies in particular. And, finally, it is crucial for the security and stability of the Mediterranean in general that Israelis and Palestinians, and Israelis and Arabs, be at last able to establish a meaningful, peaceful coexistence with one another.

NOTES

[1] Press Communiqué M-NAC-D-2 (97) 149—Meeting of the NAC in Defence Ministers Session, Brussels, December 2, 1997, paragraph 18.

[2] The phrase is used in the article by Fred Tanner, "The Euro-Med Partnership: Prospects for Arms Limitations and Confidence-Building after Malta," *The International Spectator* 32, 2 (April–June 1997): 25.

[3] NAA, Mediterranean Special Group, *Draft General Report—NATO's Role in the Mediterranean,* AP 245 GSM (97) 9 by Mr. Pedro Moya (Spain), General Rapporteur Brussels, International Secretariat, August 1997), 13. For full details of the 1997 RAND Report see, F. Stephen Larrabee et al., *NATO's Mediterranean Initiative: Policy Issues and Dilemmas* (Santa Monica, Ca.: RAND, 1998).

[4] Phil Williams, "Multilateralism: Critique and Appraisal," in *Multilateralism and Western Strategy,* Michael Brenner, ed. (New York and Basingstoke and London: St. Martin's and Macmillan, 1995), 212.

[5] John Gerard Ruggie, "Multilateralism: The Anatomy of an Institution," in *Multilateralism Matters: The Theory and Praxis of an Institutional Form,* John Gerard Ruggie, ed. (New York and Oxford: Columbia University Press, 1993), 11.

[6] Secretary-General's Remarks at the Conference on the Mediterranean Dialogue and the New NATO, Valencia, February 25, 1999.

[7] For a discussion of the RAND Corporation draft report issued in 1997, see Nicola de Santis, "The Future of NATO's Mediterranean Initiative," *NATO Review* 46, 1 (spring 1998): 32–35.

Bibliography

BOOKS AND MONOGRAPHS

Aliboni, Roberto. *European Security across the Mediterranean.* Paris: Chaillot Papers 2, Institute for Security Studies, WEU, 1991.

Aliboni, Roberto, ed. *Southern European Security in the 1990s.* London and New York: Pinter, 1992.

Aliboni, Roberto, George Joffé, and Tim Niblock, eds. *Security Challenges in the Mediterranean Region.* London and Portland, Oreg.: Frank Cass, 1996.

Ayoob, Mohammed. *The Third World Security Predicament—State Making, Regional Conflict and the International System.* Boulder, Colo., and London: Lynne Rienner, 1995.

Booth, Ken, ed. *New Thinking about Strategy and International Security.* London: Harper Collins Academic, 1991.

Braudel, Fernand. *The Mediterranean and the Mediterranean World in the Age of Philip II.* New York: Harper and Row, 1972.

Brenner, Michael, ed. *Multilateralism and Western Strategy.* New York and Basingstoke and London: St. Martin's and Macmillan, 1995.

Buzan, Barry. *People, States and Fear: The National Security Problem in International Relations.* Chapel Hill: University of North Carolina Press, 1983.

Calleya, Stephen C. *Navigating Regional Dynamics in the Post–Cold War World: Patterns of Relations in the Mediterranean Area.* Aldershot, and Brookfield, Vt.: Dartmouth, 1997.

Cantori, Louis J., and Steven L. Spiegel. *The International Politics of Regions: A Comparative Approach.* Englewood Cliffs, N.J.: Prentice-Hall, 1990.

Chipman, John, ed. *NATO's Southern Allies: Internal and External Challenges*. New York and London: Routledge, 1988.

Coffey, Joseph I., and Gianni Bonvicini, eds. *The Atlantic Alliance and the Middle East*. Basingstoke and London: Macmillan, 1989.

Collinson, Sarah. *Shore to Shore—The Politics of Migration in Euro-Maghreb Relations*. London, and Washington, D.C.: The Royal Institute of International Affairs and The Brookings Institution, 1996.

Cook, Don. *Forging the Alliance: NATO 1945 to 1950*. London: Secker and Warburg, 1989.

Desjardins, Marie-France. *Rethinking Confidence-Building Measures: Obstacles to Agreement and the Risks of Overselling the Process*. Oxford and New York: Adelphi Paper 307, Oxford University Press for the International Institute for Strategic Studies, 1996.

Deutsch, Karl W. *Political Community and the North Atlantic Area*. Princeton, N.J.: Princeton University Press, 1957.

Echeverria, Carlos. *Cooperation in Peacekeeping among the Euro-Mediterranean Armed Forces*. Paris: Chaillot Papers 35, Institute for Security Studies, WEU, 1999.

Faria, Fernanda, and Alvaro Vasconcelos. *Security in Northern Africa: Ambiguity and Reality*. Paris: Chaillot Papers 25, Institute for Security Studies, WEU, 1996.

Flanagan, Stephen J., and Fen Osler Hampson, eds. *Securing Europe's Future*. London and Sydney: Croom Helm, 1986.

Freedman, Lawrence, ed. *The Troubled Alliance: Atlantic Relations in the 1980s*. London: Heinemann, 1993.

Gillespie, Richard, ed. *Mediterranean Politics, Vol.2*. London: Pinter, 1996.

———. *Mediterranean Politics, Vol.1*. London: Pinter, 1994.

Golden, James R., Daniel J. Kaufman, Asa A. Clark IV, and David H. Patraeus, eds. *NATO at Forty: Change, Continuity and Prospects*. Boulder, Colo., and London: Westview, 1989.

Gulliver, P. H. *Disputes and Negotiations: A Cross-Cultural Perspective*. New York: Academic Press, 1979.

Hermann, Charles F., Charles W. Kegley, and James N. Rosenau, eds. *New Directions in the Study of Foreign Policy*. Boston: Allen and Unwin, 1987.

Holmes, John W., ed. *Maelstrom: The United States, Southern Europe and the Challenges of the Mediterranean*. Cambridge, Mass. The World Peace Foundation, 1995.

Inbar, Efraim, and Shmuel Sandler, eds. *Middle Eastern Security: Prospects for an Arms Control Regime*. London and Portland, Oreg.: Frank Cass, 1995.

Janning, Josef, and Dirk Romberg, eds. *Peace and Stability in the Middle East and North Africa*. Gütersloh: Bertelsmann Foundation, 1996.

Jawad, Haifaa A., ed. *The Middle East in the New World Order*. 2d ed. Basingstoke and London: Macmillan, 1997.

Jopp, Matthias, ed. *The Implications of the Yugoslav Crisis for Western Europe's Foreign Relations*. Paris: Chaillot Papers 17, Institute for Security Studies, WEU, 1994.

Kaplan, Lawrence S. *NATO and the United States: The Enduring Alliance*. Boston: Twayne, 1988.

Keohane, Robert O. *After Hegemony: Cooperation and Discord in the World Political Economy*. Princeton, N.J.: Princeton University Press, 1984.

Krasner, Stephen D., ed. *International Regimes*. Ithaca, N.Y., and London: Cornell University Press, 1983.

Kupchan, Charles A. *The Persian Gulf and the West: The Dilemmas of Security*. Winchester, Mass.: Allen and Unwin, 1987.

Larrabee, F. Stephen, Jerrold Green, Ian O. Lesser, and Michele Zanini. *NATO's Mediterranean Initiative: Policy Issues and Dilemmas*. Santa Monica, Calif.: RAND, 1998.

Lenzo, Guido, and Laurence Martin, eds. *The European Security Space*. Paris: Institute for Security Studies, WEU, 1996.

Lesser, Ian O. *Mediterranean Security: New Perspectives and Implications for US Policy,* Santa Monica, Calif.: RAND, 1992.

Lister, Marjorie. *The European Union and the South: Relations with Developing Countries*. London and New York: Routledge, 1997.

Lund, Michael S. *Preventing Violent Conflicts—A Strategy for Peacetime Diplomacy*. Washington, D.C.: US Institute of Peace, 1996.

Mahncke, Dieter. *Parameters of European Security*. Paris: Chaillot Papers 10, Institute for Security Studies, WEU, 1993.

Maoz, Zeev, ed. *Regional Security in the Middle East: Past, Present and Future*. London and Portland, Oreg.: Frank Cass, 1997.

Navias, Martin. *Going Ballistic—The Build-Up of Missiles in the Middle East*. London: Brassey's, 1993.

Owen, Roger. *State, Power and Politics in the Making of the Modern Middle East*. London and New York: Routledge, 1992.

Payne, Richard J. *The West European Allies, the Third World and US Foreign Policy: Post–Cold War Challenges*. New York, Westport, Conn., and London: Greenwood, 1991.

Peters, Joel. *Pathways to Peace: The Multilateral Arab-Israeli Peace Talks*. London: The Royal Institute of International Affairs, 1996.

Pierre, Andrew J., and William B. Quandt. *The Algerian Crisis: Policy Options for the West*. Washington, D.C.: The Brookings Institution, The Carnegie Endowment for International Peace, 1996.

Platt, Alan, ed. *Arms Control and Confidence-Building in the Middle East*. Washington, D.C.: US Institute of Peace, 1992.

Ruggie, John Gerard, ed. *Multilateralism Matters: The Theory and Praxis of an Institutional Form*. New York and Oxford: Columbia University Press, 1993.

Sandole, Dennis J. D., and Hugo van der Merwe, eds. *Conflict Resolution Theory and Practice: Integration and Application*. Manchester and New York: Manchester University Press, 1993.

Snyder, Jed F. *Defending the Fringe—NATO, the Mediterranean and the Persian Gulf*. Boulder, Colo.: Westview, 1987.

Stenhouse, Mark, and Bruce George. *NATO and Mediterranean Security: The New Central Region*. London: London Defence Studies 22, Brassey's for The Centre for Defence Studies, 1994.

Stuart, Douglas T., and William Tow. *The Limits of Alliance: NATO's Out-of-Area Problems since 1949*. Baltimore, Md., and London: Johns Hopkins University Press, 1990.

Thornton, Thomas P. *The Challenges to US Policy in the Third World: Global Responsibilities and Regional Devolution*. Boulder, Colo.: Westview, 1986.

Waever, Ole, Barry Buzan, Morton Kelstruf, and Pierre Lemaitre. eds. *Identity, Migration and the New Security Agenda in Europe*. New York: St. Martin's, 1993.

Wriggins, W. Howard, ed. *Dynamics of Regional Politics: Four Systems of the Indian Ocean Rim*. New York: Columbia University Press, 1992.

The Military Balance 1994–95. London: Brassey's for the International Institute for Strategic Studies, 1994.

The World in 1999. London: The Economist Group, 1998.

ARTICLES AND NOTES

Aliboni, Roberto. "Confidence-Building, Conflict Prevention and Arms Control in the Euro-Mediterranean Partnership." *Perceptions* 2, 4 (December 1997–February 1998): 73–86.

Alkadiri, Raad. "Profile: Jordan's Fading Democratic Façade." *Mediterranean Politics* 3, 1 (summer 1998): 170–75.

Asmus, Ronald D., Stephen Larrabee, and Ian O. Lesser. "Mediterranean Security: New Challenges, New Tasks." *NATO Review* 44, 3 (May 1996): 25–31.

Barbé, Esther. "The Barcelona Conference: Launching Pad of a Process." *Mediterranean Politics* 1, 1 (summer 1996): 25–42.

Bin, Alberto. "Strengthening Cooperation in the Mediterranean: NATO's Contribution." *NATO Review* 46 4 (winter 1998): 24–27.

Blunden, Margaret. "Insecurity on Europe's Southern Flank." *Survival* 36, 2 (summer 1994): 134–48.

Booth, Ken, and Peter Vale. "Security in Southern Africa: After Apartheid, Beyond Realism." *International Affairs* 71, 2 (April 1995): 285–304.

Buzan, Barry. "New Patterns of Global Security in the Twenty-First Century." *International Affairs* 67, 3 (July 1991): 431–51.

Calleya, Stephen C. "The Euro-Mediterranean Process after Malta: What Prospects?" *Mediterranean Politics* 2, 2 (autumn 1997): 1–22.

———. "Post–Cold War Regional Dynamics in the Mediterranean Area." *Mediterranean Quarterly* 7, 3 (summer 1996): 42–54.

Carter, Ashton B., and David B. Ormond. "Countering the Proliferation Risks: Adapting the Alliance to the New Security Environment." *NATO Review* 44, 5 (September 1996): 10–15.

Cassandra. "The Impending Crisis in Egypt." *Middle East Journal* 49, 1 (winter 1995): 9–27.

Cordenas, Sonia C. "The Contested Territories of Ceuta and Melilla." *Mediterranean Quarterly* 7, 1 (winter 1996): 118–31.

Dabrowska, Karen. "The Mediterranean: Cradle of Armed Conflict." *Dialogue* (June 1997), 3.

Djebbar, Saad. "The Algerian Presidential Election and Its Consequences." *Mediterranean Politics* 1, 1 (summer 1996): 118–23.

Elgstrom, Ole. "National Culture and International Negotiations." *Cooperation and Conflict* 29, 3 (1994): 289–301.

Estrella, Rafael. "EUROFOR-EUROMARFOR—A Brief Note." Available from the NAA Secretariat, Brussels.

Evans, Gareth. "Cooperative Security and Intra-State Conflict." *Foreign Policy* 96 (fall 1994): 3–20.

Faria, Fernanda. "The Mediterranean: A New Priority in Portuguese Foreign Policy." *Mediterranean Politics* 1, 2 (autumn 1996): 212–30.

Fenech, Dominic. "The Relevance of European Security Structures to the Mediterranean (and Vice Versa)." *Mediterranean Politics* 2, 1 (summer 1997): 149–76.

Galduf, Josep M. Jordan. "Spanish-Moroccan Economic Relations." *Mediterranean Politics* 2, 1 (summer 1997): 49–63.

Ghebali, Victor-Yves. "Toward a Mediterranean Helsinki-Type Process." *Mediterranean Quarterly* 4, 1 (winter 1993): 92–101.

Grimaud, Nicole. "Tunisia: Between Control and Liberalization." *Mediterranean Politics* 1, 1 (summer 1996): 95–106.

Haftendorn, Helga. "The Security Puzzle: Theory-Building and Discipline-Building in International Security." *International Studies Quarterly* 35, 1 (March 1991): 3–17.

Hollis, Rosemary. "Europe and the Middle East: Power by Stealth." *International Affairs* 73, 1 (January 1997): 15–29.

Huntington, Samuel P. "The Clash of Civilizations." *Foreign Affairs* 72, 3 (summer 1993): 22–49.

Hurrell, Andrew. "Latin America in the New World Order: A Regional Bloc of the Americas." *International Affairs* 48, 1 (January 1992): 121–31.

Ibrahim, Saad Eddin. "Crises, Elites and Democratization in the Arab world." *Middle East Journal* 47, 2 (spring 1993): 292–305.

Jawad, Haifaa A. "Islam and the West: How Fundamental Is the Threat?" *The RUSI Journal* 140, 4 (August 1995): 34–35.

Jentleson, Bruce. "The Middle East Arms Control and Regional Security (ACRS) Talks: Progress, Problems and Prospects." *Institute on Global Conflict and Cooperation, University of California,* Policy Paper 26 (September 1996).

Joffé, George. "Southern Attitudes towards an Integrated Mediterranean Region." *Mediterranean Politics* 2, 1 (summer 1997): 12–29.

Johnston, Alastair Ian. "Thinking about Strategic Culture." *International Security* 19 (spring 1995): 32–64.

Jones, Peter M. "New Directions in Middle East Deterrence: Implications for Arms Control." *Middle East Review of International Affairs e-mail Journal,* 4, 4 (December 1997).

Kienle, Eberhard. "Destabilization through Partnership? Euro-Mediterranean Relations after the Barcelona Declaration." *Mediterranean Politics* 3, 2 (autumn 1998): 1–20.

Kukat, Yulağ Tekin. "Turkey's Entry to the North Atlantic Treaty Organization." *Diş Politika* (Ankara) 10, 3–4 (1983): 50–77.

Leveau, Rémy. "Morocco at the Crossroads." *Mediterranean Politics* 2, 2 (autumn 1997): 95–117.

Lewis, William H. "Algeria at 35: The Politics of Violence." *The Washington Quarterly* 19, 3 (summer 1996): 3–18.

Loescher, Gil. "The European Community and Refugees." *International Affairs* 65, 4 (autumn 1989): 617–36.

Mangold, Peter. "Security: New Ideas, Old Ambiguities." *The World Today* 47, 2 (February 1991): 30–32.

Marks, Jon. "High Hopes and Low Motives: The New Euro-Mediterranean Partnership Initiative." *Mediterranean Politics* 1, 1 (summer 1996): 1–24.

Mazarr, Michael J. "Culture and International Relations: A Review Essay." *The Washington Quarterly* 19, 2 (spring 1996): 177–97.

Murphy, Emma C. "Ten Years On—Ben Ali's Tunisia." *Mediterranean Politics* 2, 3 (winter 1997): 114–22.

Neumann, Iver B. "A Region-Building Approach to Northern Europe." *Review of International Studies* 20, 1 (January 1994): 53–74.

Nordam, Jette. "The Mediterranean Dialogue: Dispelling Misconceptions and Building Confidence." *NATO Review* 45, 4 (July–August 1997): 26–29.

Norton, Augustus Richard. "The Future of Civil Society in the Middle East." *Middle East Journal* 47, 2 (spring 1993): 205–16.

Payne, Keith B. "Post–Cold War Deterrence and Missile Defense." *Orbis* 39, 2 (spring 1995): 201–23.

Pazzanita, Anthony G. "Political Transition in Mauritania: Problems and Prospects." *Middle East Journal* 53, 1 (winter 1999): 44–58.

Perthes, Volker. "Security Perceptions and Cooperation in the Middle East: The Political Dimension." *The International Spectator* 31, 4 (October–December 1996): 53–62.

Putnam, Robert D. "Diplomacy and Domestic Politics: The Logic of Two-Level Games." *International Organization* 42, 3 (summer 1988): 427–60.

Richards, Alan. "Economic Imperatives and Political Systems." *Middle East Journal* 47, 2 (spring 1993): 217–27.

Roberts, Hugh. "Algeria: A Controversial Constitution." *Mediterranean Politics* 2, 1 (summer 1997): 188–92.

———. "Algeria's Ruinous Impasse and the Honourable Way Out." *International Affairs* 71, 2 (April 1995): 247–67.

Robinson, Glen E. "Can Islamists Be Democrats? The Case of Jordan." *Middle East Journal* 51, 3 (summer 1997): 373–87.

Santis, Nicola de. "The Future of NATO's Mediterranean Initiative." *NATO Review* 46, 1 (spring 1998): 32–35.

Schulte, Gregory L. "Responding to Proliferation—NATO's Role." *NATO Review* 43, 4 (July 1995): 15–19.

Serfaty, Simon. "Algeria Unhinged: What Next? Who Cares? Who Leads?" *Survival* 38, 4 (winter 1996–97): 137–53.

Sloan, Stanley. "US Perspectives on NATO's Future." *International Affairs* 31, 2 (April 1995): 217–31.

Solingen, Etel. "The Domestic Sources of Regional Regimes: The Evolution of Nuclear Ambiguity in the Middle East." *International Studies Quarterly* 38, 2 (June 1994): 305–37.

Spencer, Claire. "Building Confidence in the Mediterranean." *Mediterranean Politics* 2, 2 (autumn 1997): 23–48.

---. "A Tale of Two Cities." *The World Today* 53, 3 (March 1997): 79–82.
Steinberg, Gerald M. "European Security and the Middle East Peace Process." *Mediterranean Quarterly* 7, 1 (winter 1996): 65–80.
Tanner, Fred. "The Euro-Med Partnership: Prospects for Arms Limitations and Confidence-Building after Malta." *The International Spectator* 32, 2 (April–June 1997): 3–25.
---. "The Mediterranean Pact: A Framework for Soft Security Cooperation." *Perceptions* 1, 4 (December 1996–February 1997): 56–67.
Terterov, Murat. "Egypt: The Islamist Challenge." *Mediterranean Politics* 1, 2 (autumn 1996): 243–49.
Waever, Ole. "Nordic Nostalgia: Northern Europe after the Cold War." *International Affairs* 68, 1 (January 1992): 77–102.
Widgren, Jonas. "International Migration and Regional Stability." *International Affairs* 66, 4 (October 1990): 749–66.
Winrow, Gareth M. "NATO and Out-of-Area: A Post–Cold War Challenge." *European Security* 3, 4 (winter 1994): 617–38.
---. "NATO and the Out-of-Area Issue: The Positions of Turkey and Italy." *Il Politico* (Pavia) 58, 4 (1993): 631–52.
Yahia, Habib Ben. "Security and Stability in the Mediterranean: Regional and International Challenges." *Mediterranean Quarterly* 4, 1 (winter 1993): 1–10.
Zielonka, Jan. "Europe's Security: A Great Confusion." *International Affairs* 67, 1 (January 1991): 127–37.

OFFICIAL DOCUMENTS AND REPORTS, AND COLLECTIONS OF DOCUMENTS AND SPEECHES

Alliance Policy Framework on Weapons of Mass Destruction—Issued at the Ministerial Meeting of the North Atlantic Council in Istanbul, 9 June 1994. Brussels, NATO Office of Information and Press, 1994.

Assembly of Western European Union. Doc. 1485, 6 November 1995. "Parliamentary Cooperation in the Mediterranean." Report submitted on behalf of the Committee for Parliamentary and Public Relations by Mr. Kitsonis, Rapporteur.

---. *Doc. 1468, 12 June 1995.* "European Armed Forces." Report submitted on behalf of the Defence Committee by Mr. de Decker, Rapporteur.

---. *Proceedings, 42nd Session, December 1996, I. Assembly Documents* (Paris: WEU), Doc. 1543, November 4,1996. "Security in the Mediterranean Region." Report submitted on behalf of the Political Committee by Mr. de Lipkowski, Rapporteur.

Barcelona Declaration Adopted at the Euro-Mediterranean Conference, 28 November 1995. Barcelona: 1995.

Boutros-Ghali, Boutros. *An Agenda for Peace—Preventive Diplomacy, Peacemaking and Peace-keeping—Report of the Secretary-General pursuant to the Statement adopted by the Summit Meeting of the Security Council on 31 January 1992*. New York: United Nations, 1992.

British Foreign Office General Correspondence Documents.

Chairman's Formal Conclusions of the Third Euro-Ministerial Conference, Stuttgart, 15–16 April 1999. Stuttgart: 1999.

Change and Continuity in the North Atlantic Alliance—Speeches by the Secretary-General of NATO, Manfred Wörner. Brussels: NATO Office of Information and Press, 1990.

CSCE, *Helsinki Final Act*. Helsinki: 1975.

———. *Budapest Document, December 1994—Budapest Summit Declaration—Towards a Common Partnership in a New Era*. Budapest: 1994.

———. *Helsinki Document—The Challenges of Change, July 1992*. Helsinki: 1992.

———. *Report of the Meeting on the Mediterranean of the Conference on Security and Cooperation in Europe, Foreseen by the Concluding Document of the Vienna Meeting 1986—Palma de Mallorca, 19 October 1990*. Palma de Mallorca: 1990.

CSCE Communication. Prague.

CSO/Journal. Prague.

London Declaration on a Transformed North Atlantic Alliance—Issued by the Heads of State and Government Participating in the Meeting of the North Atlantic Council in London, 5–6 July 1990. Brussels: NATO Office of Information and Press, 1990.

Mediterranean Forum—Summary Report—Meetings of High Officials, Sainte-Maxime, 8–9 April 1995.

NAA Civilian Affairs Committee, Sub-Committee on the Mediterranean Basin. *Draft Interim Report—Cooperation for Security in the Mediterranean: NATO and EU Contributions*, AM 83 CC/MB (96) 1 by Mr. Pedro Moya (Spain), Rapporteur. Brussels: International Secretariat, May 1996.

———. Civilian Affairs Committee, Sub-Committee on the Mediterranean Basin. *Report—Frameworks for Cooperation in the Mediterranean*, AM 259 CC/MB (95) 7 by Mr. Pedro Moya (Spain), Rapporteur. Brussels: International Secretariat, October 1995.

———. Civilian Affairs Committee, Sub-Committee on the Mediterranean Basin/Sub-Committee on the Southern Region. *Report—Tunis, Tunisia, 6–8*

June 1995, AM 179 rev.1 CC/MB (95) 6. Brussels: International Secretariat, September 1995.

———. Civilian Affairs Committee, Sub-Committee on the Mediterranean Basin. *The Rise of Islamic Radicalism and the Future of Democracy in North Africa,* AL 199 CC/MB (94), 4 by Mr. Pedro Moya (Spain), Rapporteur. Brussels: International Secretariat, November 1994.

———. Civilian Affairs Committee, Sub-Committee on the Mediterranean Basin. *Report—Seminar on the Maghreb, Paris, 9–10 June 1994,* AL 155 CC/MB (94) 3. Brussels: International Secretariat, July 1994.

———. Civilian Affairs Committee, Sub-Committee on the Mediterranean Basin. *Interim Report,* AJ 236 CC/MB (92) 4 by Mr. Augusto Borderas (Spain), Rapporteur. Brussels: International Secretariat, November 1992.

———. Civilian Affairs Committee, Sub-Committee on the Mediterranean Basin. *Interim Report,* AI 236 CC/MB (91) 6 by Mr. Augusto Borderas (Spain), Rapporteur. Brussels: International Secretariat, October 1991.

———. Mediterranean Special Group. *Draft General Report—NATO's Role in the Mediterranean,* AP 245 GSM (97) 9 by Mr. Pedro Moya (Spain), General Rapporteur. Brussels: International Secretariat, August 1997.

———. Mediterranean Special Group. *Draft Interim Report—NATO's Role in the Mediterranean,* AP 115 GSM (97) 5 by Mr. Pedro Moya (Spain), Rapporteur. Brussels: International Secretariat, April 1997.

———. Mediterranean Special Group. *Chairman's Conclusions and 1997 Work Programme, 4 December 1996, Lisbon, Portugal,* AN 319 GSM (96) 1. Brussels: International Secretariat, December 1996.

———. Political Committee. *Continental Drift,* AL 221 PC (94) 5 by Mr. Bruce George (United Kingdom), Rapporteur. Brussels: International Secretariat, November 1994.

———. Political Committee. *Europe and Transatlantic Security in a Revolutionary Age,* AK 244 PC (93) 6 by Mr. Bruce George (United Kingdom), General Rapporteur. Brussels: International Secretariat, October 1993.

———. Political Committee. *NATO and the New Arc of Crisis: Dialectics of Russian Foreign Policy,* AJ 260 PC (92) 5 by Mr. Bruce George (United Kingdom), Rapporteur. Brussels: International Secretariat, November 1992.

———. Political Committee, Sub-Committee on Out-of-Area Security Challenges to the Alliance. *Interim Report,* AC 182 PC/OA (85) 2 by Mr. Herrero de Minon (Spain), Rapporteur. Brussels: International Secretariat, October 1985.

———. Political Committee, Sub-Committee on the Southern Region. *Draft Interim Report,* AM 106 PC/SR (95) 1 by Mr. Rodrigo de Rato (Spain), Rapporteur. Brussels: International Secretariat, May 1995.

———. Political Committee, Sub-Committee on the Southern Region. *Cooperation and Security in the Mediterranean,* AL 223 PC/SR (94) 2 by Mr. Rodrigo de Rato (Spain), Rapporteur. Brussels: International Secretariat, November 1994.

———. Political Committee, Sub-Committee on the Southern Region. *Interim Report,* AJ 262 PC/SR (92) 4 by Mr. Miguel Herrero (Spain), Rapporteur. Brussels: International Secretariat, November 1992.

———. Presidential Task Force. *Final Report—America and Europe: The Future of NATO and the Transatlantic Relationship.* Co-Chairmen Loic Bouvard (France) and Charlie Rose (United States). Brussels: International Secretariat, September 1993.

———. Standing Committee. *Restructuring NAA Mediterranean Activities,* AN 33 SC (96) 9 by Mr. Pedro Moya (Spain), Rapporteur of the Sub-Committee on the Mediterranean Basin and Mr. Anders Sjaastad (Norway), Chairman of the Defence and Security Committee. Brussels: International Secretariat, March 1996.

NATO Letter. Vol.5, Special Supplement to no.1, January 1, 1957. Non-Military Cooperation in NATO—Text of the Report of the Committee of the Three.

NATO to the Year 2000: Challenges for Coalition Deterrence and Defense. Report of the Working Group of the Atlantic Council of the United States on the future of NATO, Washington, D.C., March 1988. Chairman, Andrew J. Goodpaster, Rapporteur, Ian O. Lesser.

OSCE Permanent Council, PC. DEC/94 5 December 1995.

———. *Permanent Council, PC. DEC/233 11 June 1998.*

———. *Ref MC (6) DEC/5 19 December 1997—MC (6) Journal No.2.*

———. *Ref MC/11/95 30 November 1995—OSCE—The Secretary General, Annual Report 1995 on OSCE Activities.*

———. *Ref. MC/22/95 7 December 1995—Statement of Egypt Presented to the Fifth Ministerial Meeting of the OSCE, Budapest, 7–8 December 1995.*

———. *Ref. SEC/288/95 16 October 1995 (Vienna)—OSCE Seminar on the OSCE Experience in the Field of Confidence-Building (Consolidated Summary), Cairo, 26–28 September 1995.*

Texts of Final Communiqués 1949–74. Brussels: NATO Information Service.

Texts of Final Communiqués 1981–85, Vol.3. Brussels: NATO Information Service.

Texts of Final Communiqués 1986–90, Vol.4. Brussels: NATO Office of Information and Press.

The Alliance's Strategic Concept—Agreed by the Heads of States and Governments Participating in the Meeting of the North Atlantic Council in Rome, 7–8 November 1991. Brussels: NATO Office of Information and Press, 1991.

Treaty on European Union. Luxembourg: Office for Official Publications of the European Communities, 1992.

WEU, *Council of Ministers, Lisbon Declaration,* Lisbon, 15 May 1995, Doc.1455.

———. *Council of Ministers, Petersberg Declaration,* Bonn, 19 June 1992.

———. *Ministerial Council, WEU Council of Ministers, Rhodes Declaration,* Rhodes, 15 May 1998, Doc.1612.

———. *Ministerial Council, WEU Council of Ministers, Erfurt Declaration,* Erfurt, 18 November 1997.

———. *Ministerial Council, WEU Council of Ministers, Declaration of WEU on the Role of WEU and Its Relations with the EU and with the Atlantic Alliance,* Amsterdam, 22 July 1997.

———. *Ministerial Council, WEU Council of Ministers, Paris Declaration,* Paris, 13 May 1997.

———. *Ministerial Council, WEU Council of Ministers, Birmingham Declaration,* Birmingham, 7 May 1996.

WEU Related Texts Adopted at EC Summit, Maastricht—10 December 1991. London: WEU Press and Information Section, December 10, 1991.

Western European Union: The Reactivation of WEU: Statements and Communiqués 1984 to 1987. London: Secretariat-General of WEU, 1988.

46th Annual Report of WEU Council to the Assembly of WEU (1 July–31 December 1994).

CONFERENCE AND WORKSHOP PAPERS AND PRESENTATIONS

Colas, Alejandro."The Limits of Mediterranean Partnership: Civil Society and the Barcelona Conference of 1995." Paper presented at the International Studies Association Convention, Toronto, March 19–22, 1997.

Gunnarsdottir, Gréta."NATO's Mediterranean Dialogue: Current Status—Future Prospects." Paper presented at the Mediterranean Dialogue Seminar organized by the NAA Mediterranean Special Group, Lisbon, December 5–6, 1996.

Lesser, Ian O. "Southern Europe and the Maghreb: US Interests and Policy Perspectives." Paper elaborating on remarks delivered at the conference Employment, Economic Development and Migration: European and US Perspectives, organized by the Luso-American Development Foundation, Lisbon, April 28, 1995.

Monteiro, Ambassador Antonio."NATO and the Mediterranean Security Dialogue (A PfP-Inspired Model)." Paper presented at the Mediterranean Dialogue Seminar organized by the NAA Mediterranean Special Group, Lisbon, December 5–6, 1996.

Silvestri, Stefano. Presentation in a NAA Workshop on Confidence-Building Measures in the Mediterranean Region, Naples, April 11–12, 1997.

Vetschera, Heinz. Presentation in a NAA Workshop on Confidence-Building Measures in the Mediterranean Region, Naples, April 11–12, 1997.

———. "Regional Security and Arms Control—The CSCE Experience with Confidence-Building Measures and Crisis Mechanisms." Paper available at the Secretariat of the NAA, Brussels.

Report of the Sixth International Antalya Conference on Security and Cooperation. Organized by the Atlantic Council of Turkey, Antalya, November 2–5, 1995.

NEWSLETTERS, NEWS SERVICES, NEWSPAPERS, MAGAZINES AND DIGESTS

Al Ahram
The Economist
Euro-Mediterranean Partnership—Monthly Information Notes
Foreign Broadcast Information Service
Guardian Weekly
Institute for Security Studies, WEU Newsletter
Keesings Record of World Events
Middle East International
NATO Review
OSCE Newsletter
The Times
La Tribune
Turkish Daily News

Index

Abdullah, King, 73, 87, 91
Achille Lauro affair, 56, 62
ACRS (Working Group on Arms
  Control and Regional Security),
  40, 110, 126–128, 159, 208, 227
  (*see also* Barcelona Process)
  and NATO's Mediterranean Initiative, 174, 189, 226
AFTA (Arab Free Trade Area),
  107–108, 210
AIS (Islamic Salvation Army), 81
Algeria, 55, 93, 107, 124, 137–138
  (*see also* CSCE, CSCM, Egypt,
  EU, France, Italy, Mauritania,
  Mediterranean Forum, Morocco, NAA-Mediterranean
  Dialogue, OSCE-Mediterranean
  Dialogue, Portugal, Soviet
  Union, Spain, Tunisia, U.S.,
  Western Mediterranean Forum,
  WEU-Mediterranean Dialogue)
  AMU, relations with, 107, 112,
    133, 150, 190
  Berbers in, 79, 90–91
  economic situation in, 81, 90, 98,
    103, 104
  EC, relations with, 107, 151
  military capability of, 109
  NATO's Mediterranean Initiative,
    relations with, xiii, 20, 72, 81,
    93, 159, 169, 170–171, 190,
    202, 229
  political situation in, 72, 75–76,
    79, 80, 81–82, 85, 90–91
  problem of migration with, 98–99,
    100, 101, 102
  role of military in, 86
Aliel-Harani, Mohammed, 82
AMU (Arab Maghreb Union), 107,
  111–112, 133, 144, 210, 227
  (*see also* Algeria, Egypt, Libya,
  Mauritania, Morocco, Tunisia,
  Western Mediterranean Forum)
Anan, Kofi, 82
Andreotti, Guilio, 134
Arab League, 16, 61–62, 201
Arafat, Yasser, xi, 13, 14, 15
Arens, Moshe, xii
Austria, 132

Balanzino, Sergio, 175
Barak, Ehud, xi-xii
Barcelona Conference, 135,
  152–153, 201, 208–212

Barcelona Process, 2, 7–8, 12–13, 41, 66, 87, 108, 171–172, 207, 208–218, 224–225, 227 (*see also* CBM, CSBM, preventive diplomacy)
  and ACRS, 215
  Britain, relations with, 209
  Egypt, relations with, 214
  France, relations with, 186, 209
  Iran, relations with, 42, 215
  Iraq, relations with, 42, 215
  Israel, relations with, 209, 214, 216
  Lebanon, relations with, 209
  Libya, relations with, 42, 209, 215
  Mauritania, relations with, 209
  Mediterranean Forum, relations with, 212–213, 217
  and Middle East peace process, xii, 13, 17, 209, 210, 213–214, 216
  NATO's Mediterranean Initiative, relations with, 177–178, 197, 199, 208, 217, 229, 230, 231
  OSCE, relations with, 217, 218
  problem of migration with, 103, 212, 213, 215
  Syria, relations with, 209
  Turkey, relations with, 214
  U.S., relations with, 9, 209, 216, 217
  WEU–Mediterranean Dialogue, relations with, xiv, 201, 202, 217
  and WMD, 213–214, 215, 216
Basri, Driss, 74, 78, 92
Belgium
  NATO, relations with, 52
  problem of migration with, 99
Ben Ali, Zine el Abidine, 61, 74–75, 77, 82, 85, 87, 91

Black Sea Economic Cooperation, 133
Bosnia, 4, 90, 99, 124 (*see also* WEU)
  NATO, relations with, xii, 113, 122, 123–124, 128, 149, 174, 181–182, 189
Bourguiba, Habib, 74, 82
Bouteflika, Abdelaziz, 76, 229
Britain, 59 (*see also* Barcelona Process, CSCM, EU, US)
  Egypt, relations with, 53, 57–58, 64
  Iraq, relations with, 90
  Mediterranean, relations with, 52, 53, 59, 110
  NATO, relations with, 52, 53, 54
  NATO's Mediterranean Initiative, relations with, 186
Bulgaria, 186

Canada
  NATO, relations with, 52, 131
  NATO's Mediterranean Initiative, relations with, 154, 168, 186
Canary Islands, 123, 170
Carter, Jimmy, 59
CBM (Confidence-building measure)
  and Barcelona Process, 213–216, 217
  and CSCE, 37–38, 41–42, 147
  definition of, 17, 36–41, 192
  and Middle East peace process, 40–42, 125–127, 159
  and NATO's Mediterranean Initiative, 2, 36–37, 45, 108, 115, 116, 168, 192, 230, 231
  and OSCE, 41–42
  and OSCE-Mediterranean Dialogue, 147, 205–206, 207
CENTO (Central Treaty Organization), 53

Central European Initiative, 132
Ceuta, 63, 92, 117n, 132
Chad, 4 (*see also* France, Libya)
Chirac, Jacques, 187, 216
Claes, Willy, 157–158, 172, 173–174
Conflict prevention (*see* preventive diplomacy)
Contadora Group, 133
Corcione, Domenico, 184
Council of Europe, 8, 144, 146, 148
Council of the Mediterranean, 8, 146, 148
Crete, 49
Crisis management, 17
  and NATO, xii, xiii, 124, 128, 129, 131, 142–143, 159
  and NATO's Mediterranean Initiative, 115, 178, 182, 184, 187, 230
Croatia, 124 (*see also* Mediterranean Forum)
CSBM (Confidence- and security-building measure):
  and Barcelona Process, 213–216, 217
  and CSCE, 38, 41–42
  definition of, 37, 38–39
  and Middle East peace process, 40–42, 126, 159
  and NATO's Mediterranean Initiative, 115, 116
  and OSCE-Mediterranean Dialogue, 205
CSCE (Conference on Security and Cooperation in Europe), 37, 126, 209 (*see also* CBM, CSBM, WEU-Mediterranean Dialogue)
  Algeria, relations with, 134, 146, 147
  Cyprus, relations with, 133
  Egypt, relations with, 134, 146, 148
  Israel, relations with, 134, 136
  Italy, relations with, 133
  Lebanon, relations with, 134, 146, 147
  Libya, relations with, 134, 146, 147
  Malta, relations with, 133, 146
  Mediterranean, relations with, 41–42, 133–134, 135, 136, 137, 146–148, 206, 207–208
  Morocco, relations with, 134, 146, 148
  Spain, relations with, 133
  Syria, relations with, 134, 147
  Tunisia, relations with, 134, 146
  Yugoslavia, relations with, 133
CSCM (Conference on Security and Cooperation in the Mediterranean), 3, 8, 133–137, 138, 139, 144, 147, 148, 154, 171, 207–208, 209
  Algeria, relations with, 137
  Britain, relations with, 136, 137
  EC, relations with, 137
  Egypt, relations with, 134, 136
  France, relations with, 134–135, 135–136
  Germany, relations with, 136
  Greece, relations with, 134, 185
  Israel, relations with, 137
  Italy, relations with, 3, 133, 134–135, 136
  Libya, relations with, 136, 137
  Palestinians, relations with, 135, 137
  Portugal, relations with, 134–135, 137
  Russia, relations with, 137
  Spain, relations with, 3, 133, 134–135, 137
  U.S., relations with, 135, 136, 137
  Yugoslavia, relations with, 137

CSCME (Conference on Security and Cooperation in the Middle East), 207–208
Culture:
definition of, 28–30, 224
Cutiliero, Jose, 149–150
Cyprus, 10, 49, 55–56, 132 (*see also* CSCE, EU, Mediterranean Forum, NAA-Mediterranean Dialogue, WEU-Mediterranean Dialogue)
Czech Republic
NATO, relations with, 228

de Gaulle, Charles, 58
de Marco, Guido, 146, 148
Democratization
in Mediterranean, prospects for, 26–27, 72–73, 87–93
de Michelis, Gianni, 132, 133, 134
Dialogue
definition of, 31–34

EAPC (Euro-Atlantic Partnership Council), 185, 228
and NATO's Mediterranean Initiative, 185
Eastern Mediterranean Pact, 52
EC (European Community) (*see also* Algeria, CSCM, Egypt, Israel, Jordan, Lebanon, Morocco, Syria, Tunisia)
Mediterranean, relations with, 61–62, 66, 107, 151
and Middle East peace process, 151

EEC (European Economic Community)
Mediterranean, relations with, 151
Egypt, 4, 52, 53, 54, 66, 100, 197, 124, 156, 158, 186 (*see also* Barcelona Process, Britain, CSCE, CSCM, EU, France, Israel, Mediterranean Forum, NAA, NAA-Mediterranean Dialogue, OSCE-Mediterranean Dialogue, Soviet Union, U.S., Western Mediterranean Forum)
Algeria, relations with, 172
AMU, relations with, 112
economic situation in, 90, 98, 104–105
EC, relations with, 151
Libya, relations with, 173
military capability of, 109–110
NATO, relations with, 124, 174, 181–182, 190
NATO's Mediterranean Initiative, relations with, 1, 157, 169, 178, 189–190, 192, 229, 230
political situation in, 75, 77, 80, 81, 82–83, 85, 90, 91
role of military in, 86–87
Sudan, relations with, 12, 111
el-Jihad, 82
El Mundo incident, 183, 204
el-Sadat, Anwar, 75, 82
En Nahda (Algeria), 79, 81
En Nahda (Tunisia), 82, 93
ESDI (European Security and Defense Identity), xiii, 2, 143, 149
EU (European Union): (*see also* Barcelona Conference, Barcelona Process)
Algeria, relations with, 152
Britain, relations with, 201
and CFSP (Common Foreign and Security Policy), xiv, 151, 201, 216
Cyprus, relations with, 152
Egypt, relations with, 144, 152
France, relations with, 132, 201
Israel, relations with, 152, 209
Italy, relations with, 187
Jordan, relations with, 152, 209

Lebanon, relations with, 153
Libya, relations with, 152, 153
Malta, relations with, 153
Mauritania, relations with, 153
Mediterranean, relations with, 8, 39, 98, 103, 106–108, 116, 139, 146, 151–153, 189, 209–210, 211
and Middle East peace process, 14–15, 16–17, 125, 153, 216
Morocco, relations with, 152, 153, 209
NATO, relations with, xiv, 116, 153, 217
Palestinians, relations with, 153, 216
Portugal, relations with, 187
problem of migration with, 101–102
Spain, relations with, 187
Syria, relations with, 153
Tunisia, relations with, 152, 153, 190, 209
Turkey, relations with, 153
WEU, relations with, xiv, 149, 151, 200, 201
Euro-Arab Dialogue, 61–62, 66, 134, 151
EUROFOR (European Rapid Operational Force), 115, 130, 149, 187, 202–204 (*see also* Libya, Morocco, Tunisia)
NATO, relations with, 203–204
NATO's Mediterranean Initiative, relations with, 202, 204, 217
OSCE, relations with, 203
UN, relations with, 203
EUROMARFOR (European Maritime Force), 115, 130, 149, 187, 202–204
NATO, relations with, 203–204
NATO's Mediterranean Initiative, relations with, 202, 204, 217

OSCE, relations with, 203
UN, relations with, 203

FFS (Front of Socialist Forces), 79, 90
Filali, Abdellatif, 190
FIS (Islamic Salvation Front), 79, 81, 86, 93
FLAM (Mauritanian African Liberation Forces), 86
FLN (National Liberation Front), 79
France, 59, 115 (*see also* Barcelona Process, CSCM, EU, Mediterranean Forum, U.S., Western Mediterranean Forum, WEU, WEU-Mediterranean Dialogue)
Algeria, relations with, 10, 52–53, 57, 58, 63, 64, 80, 99, 101, 102, 104, 115, 140, 171
Chad, relations with, 5, 60, 63
Egypt, relations with, 57–58, 64
Iran, relations with, 140
Iraq, relations with, 140
Libya, relations with, 5, 60, 63, 112, 140
Mediterranean, relations with, 2, 52–53, 59, 63, 110, 130, 131, 132, 149
Morocco, relations with, 52, 63, 64
NATO, relations with, 51, 52–53, 54, 58, 132, 143, 156–157, 186–187
NATO's Mediterranean Initiative, relations with, 168, 171, 183, 186–187, 231
problem of migration with, 10–11, 99, 101, 102
Tunisia, relations with, 52, 63, 64, 112, 113, 115

Gadhafi, Mu'ammar, 10, 60, 62, 63, 73, 109, 173, 183, 204

Germany (*see also* CSCM)
  Mediterranean, relations with, 60–61, 103
  NATO's Mediterranean Initiative, relations with, 186, 231
  problem of migration with, 99
Ghannouchi, Rachid, 82, 93
GIA (Armed Islamic Group), 81, 82
Gorbachev, Mikhail, 38
Greece, 71 (*see also* CSCM, Mediterranean Forum, Turkey, U.S., Western Mediterranean Forum, WEU, WEU-Mediterranean Dialogue)
  Mediterranean, relations with, 52, 53, 63, 137
  NATO, relations with, 10, 51–52, 53–54, 55–56
  NATO's Mediterranean Initiative, relations with, 185, 223
Group of Contact, 133

Hamas (Algeria), 81
Hamas (Israel), 14, 15, 16, 80
HASM (Islamic Movement of Mauritania), 84
Hassan, Crown Prince, 170
Hassan, King, 63, 74, 78, 86, 91, 92
HATM (Movement of Reform and Renewal in Morocco), 83
Helsinki Final Act (1975), 37, 41, 126, 133–134, 135
Hezbollah, 14
Hungary, 132
  NATO, relations with, 228
Hussein, King, 73–74, 77, 84, 105, 170
Hussein, Saddam, 27, 92, 105, 122, 123

IAF (Islamic Action Front), 77, 84–85

IFOR (Implementation Force), 4, 124, 174, 184, 190
IMF (International Monetary Fund), 88–89, 104, 105, 106
IPU (Inter-Parliamentary Union), 137, 200, 209
Iran, 52, 59 , 80, 127, 131, 208 (*see also* Barcelona Process, France, Israel, Russia, Turkey)
  Iraq, war with, 55, 59–60, 62
  military capability of, 109, 121
Iraq, 27, 55, 123, 127, 131, 208 (*see also* Barcelona Process, Britain, France, Iran, Israel, Jordan, Mauritania, Turkey, UN, U.S.)
  Kuwait, invasion of, 4, 12, 112, 113, 121–123, 149
  military capability of, 109
  NATO, relations with, 122, 123
Islam
  and NATO's Mediterranean Initiative, 85, 157–158, 174
Israel, 4, 108, 113, 131, 232 (*see also* Barcelona Process, CSCE, CSCM, EU, Mediterranean Forum, Middle East peace process, NAA, NAA-Mediterranean Dialogue, OSCE-Mediterranean Dialogue, Turkey, UN, U.S., WEU-Mediterranean Dialogue)
  EC, relations with, 151
  Egypt, relations with, 14, 16, 40, 61, 86–87, 109–110, 127, 174, 189
  Iran, relations with, 109
  Iraq, relations with, 123
  Jordan, relations with, 13, 16, 40, 84–85, 105, 122, 126, 170, 207
  Lebanon, relations with, xi, 13, 14, 16, 59
  military capability of, 110

NATO's Mediterranean Initiative, relations with, 1, 36, 157, 169–170, 173, 176, 178, 182, 185, 187–192 passim, 226, 227, 228–229, 230
Palestinians, relations with, xi-xii, 2, 13, 14–17, 19, 20, 72, 122, 232
political situation in, xi-xii, 14, 16, 71–72, 80, 225
Syria, relations with, xi, 13–14, 16, 40, 109
Italy, 50, 59, 115 (*see also* CSCE, CSCM, EU, Mediterranean Forum, OSCE-Mediterranean Dialogue, US, Western Mediterranean Forum, WEU, WEU-Mediterranean Dialogue)
Algeria, relations with, 104, 115, 171
Libya, relations with, 10, 62–63, 109
Mediterranean, relations with, 2, 52, 56, 59, 60, 130, 131, 132, 133, 137, 149, 168
NATO, relations with, 53, 203
NATO's Mediterranean Initiative, relations with, 155, 171, 172, 183–184, 186, 187
problem of migration with, 10–11, 79, 99, 100, 102
Tunisia, relations with, 115

Jama'at Islamiyya, 80, 82
Japan, 206
Jordan, 4, 58, 65, 124 (*see also* EU, Israel, Mediterranean Forum, NAA-Mediterranean Dialogue, OSCE-Mediterranean Dialogue, US, WEU-Mediterranean Dialogue)
EC, relations with, 151

economic situation in, 77, 90, 98, 105
Iraq, relations with, 105, 170, 190
NATO, relations with, 124, 174, 181–182, 190
NATO's Mediterranean Initiative, relations with, 1, 4, 168, 169, 170, 171, 175, 178, 190–191, 192, 229
political situation in, 73–74, 76–77, 84–85, 90, 91
role of military in, 87
Syria, relations with, 105
Juppe, Alain, 212–213
Justice and Charity association, 83

Kosovo
NATO, relations with, xii, 10, 122, 124–125, 223
Kruzel, Joseph, 139, 173, 183–184
Kuwait, 100 (*see also* Iraq)

Labour party, xi-xii
Lebanon, 58, 59, 125, 126, 127 (*see also* Barcelona Process, CSCE, EU, Israel, Morocco, NAA-Mediterranean Dialogue, U.S.)
EC, relations with, 151
NATO's Mediterranean Initiative, relations with, 169, 171, 190
Libya, 12, 55, 107, 114, 123, 127, 131, 159, 173 (*see also* Barcelona Process, CSCE, CSCM, Egypt, EU, France, Italy, Mauritania, Mediterranean Forum, Morocco, Portugal, Soviet Union, Spain, Tunisia, UN, U.S., Western Mediterranean Forum, WEU-Mediterranean Dialogue)
AMU, relations with, 107, 112, 133, 190, 209
Chad, relations with, 5, 60, 63, 109

Libya *(continued)*
  EUROFOR, relations with, 204
  military capability of, 108–109
  NATO's Mediterranean Initiative, relations with, 20, 169, 171, 173, 190, 226
Likud party, xii, 3, 13, 14, 169
Luxembourg
  NATO, relations with, 52

Macedonia, 124
Madani, Abassi, 81–82
Madeira, 123, 170
Malta, 8, 148, 158 *(see also* CSCE, EU, NAA-Mediterranean Dialogue, OSCE-Mediterranean Dialogue, Western Mediterranean Forum, WEU-Mediterranean Dialogue)
Matoub, Lounes, 91
Mauritania, 3, 4, 8, 136 *(see also* Barcelona Process, EU, Mediterranean Forum, Morocco, NAA-Mediterranean Dialogue, OSCE-Mediterranean Dialogue, Portugal, Spain, Western Mediterranean Forum, WEU-Mediterranean Dialogue)
  Algeria, relations with, 110, 190
  AMU, relations with, 107, 133, 190
  economic situation in, 98, 105, 111
  Iraq, relations with, 75, 122–123, 170
  Libya, relations with, 173, 190
  Mali, relations with, 111
  NATO's Mediterranean Initiative, relations with, 1, 157, 169, 170, 174, 182, 190
  political situation in, 75, 77, 78, 84, 85, 90, 91
  role of military in, 86, 91
  Senegal, relations with, 86, 111
  Sudan, relations with, 84
  Tunisia, relations with, 84
Mediterranean
  definition of, 3–9
Mediterranean Forum, 8, 144–146, 197, 212–213, 225 *(see also* Barcelona Process, Middle East peace process)
  Albania, relations with, 212
  Algeria, relations with, 145
  Croatia, relations with, 212
  Cyprus, relations with, 212
  Egypt, relations with, 144–145
  France, relations with, 144, 145
  Greece, relations with, 145
  Israel, relations with, 145, 212
  Italy, relations with, 144, 145
  Jordan, relations with, 212
  Libya, relations with, 145, 212
  Malta, relations with, 145, 212
  Mauritania, relations with, 212
  Morocco, relations with, 145
  NATO's Mediterranean Initiative, relations with, 116, 217
  Portugal, relations with, 145
  problem of migration with, 103
  Russia, relations with, 212
  Slovenia, relations with, 212
  Spain, relations with, 145
  Syria, relations with, 145
  Tunisia, relations with, 145
  Turkey, relations with, 145
  and WMD, 213
Mediterranean Pact, 53
Mediterranean Security Pact, 52
Melilla, 63, 92, 117n, 132
MENA (Middle East-North Africa) economic conferences, 16, 125–126, 128, 210, 211
MENAFIO (Middle East and North Africa Financial Intermediary Organization), 211
Middle East Command, 53

Middle East Defense Organization, 53
Middle East peace process, xi-xii, 3,
 11-17, 20, 40, 55, 122,
 125-128, 169, 207-208 (*see
 also* Barcelona Process, CBM,
 CSBM, EC, EU)
 and Mediterranean Forum, 145
 and NATO's Mediterranean Initiative, xii, xiv, 13, 17, 127-128,
 155, 169-170, 176, 177-178,
 192, 218, 223, 231
 and US, 2, 13, 14-15, 16-17, 127,
 144, 156, 186
Migration
 problem of, 10-11, 79, 98-103,
 114, 123, 131 (*see also* Algeria,
 Barcelona Process, Belgium,
 EU, France, Germany, Italy,
 Mediterranean Forum, Morocco, Netherlands, Portugal,
 Spain, Tunisia)
Mintoff, Dom, 133
Mitterrand, François, 63, 137
Monteiro, Antonio, 184
Morocco, 107, 124, 125, 137, 158
 (*see also* CSCE, EU, France,
 Mediterranean Forum, NAA,
 NAA-Mediterranean Dialogue,
 Portugal, Spain, U.S., Western
 Mediterranean Forum, WEU-
 Mediterranean Dialogue)
 Algeria, relations with, 11-12, 92,
 102, 104, 110-111, 190, 229
 AMU, relations with, 107, 112,
 133, 150
 Berbers in, 91
 economic situation in, 90, 98,
 105-106
 EC, relations with, 63, 107, 151
 EUROFOR, relations with, 204
 Lebanon, relations with, 190
 Libya, relations with, 63, 173,
 190
 Mauritania, relations with, 12,
 110, 111, 169, 190
 NATO, relations with, 124, 174,
 181-182, 190
 NATO's Mediterranean Initiative,
 relations with, 1, 157, 169, 174,
 181, 190, 229
 political situation in, 74, 78, 81,
 83, 85, 90, 91-92
 problem of migration with, 98-99,
 101, 102
 role of military in, 86
 Syria, relations with, 190
MPS (Islamist Movement for a
 Peaceful Society), 79, 81
MTI (Islamic Tendency Movement),
 82
Mubarak, Hosni, 75, 77, 87, 111,
 144-145, 204
Muslim Brotherhood (Egypt), 82-83
Muslim Brotherhood (Jordan), 84

NAA (North Atlantic Assembly),
 153, 155 (*see also* Egypt, Israel,
 Morocco)
NAA-Mediterranean Dialogue,
 153-154, 197, 198
 Algeria, relations with, 153
 Cyprus, relations with, 153, 198
 Egypt, relations with, 153, 198
 Israel, relations with, 153, 198
 Jordan, relations with, 153, 198
 Lebanon, relations with, 153
 Malta, relations with, 153
 Mauritania, relations with, 153,
 198
 Morocco, relations with, 153, 198
 NATO's Mediterranean Initiative,
 relations with, 198, 217
 Palestinians, relations with, 153,
 198
 Syria, relations with, 153
 Tunisia, relations with, 153, 198

NACC (North Atlantic Cooperation Council), 128, 142, 143, 185, 228
and Mediterranean, 153–154, 155, 228
Nahnah, Mahfoud, 81
Nasser, Gamal Abdel, 54
NATO (North Atlantic Treaty Organization) and Ad Hoc Group on the Mediterranean, 54–55, 157, 179
NATO and CCMS (Committee on the Challenges of Modern Society), and NATO's Mediterranean Initiative, 177
NATO and Expert Working Group on the Middle East and the Maghreb, 54–55, 179
NATO and Harmel Report, 44–45, 58, 142
NATO and International Staff, and NATO's Mediterranean Initiative, 35, 36, 168, 172–173, 174, 175–176, 178, 179, 180
NATO and MCG (Mediterranean Cooperation Group), and NATO's Mediterranean Initiative, 1, 35, 168, 174, 178–180, 182, 183, 226
NATO and Military Committee, and NATO's Mediterranean Initiative, 181
NATO and NAC (North Atlantic Council), and NATO's Mediterranean Initiative, 35, 36, 172, 174, 175, 178–179, 184
NATO and NAVOCFORMED (Naval On-Call Force Mediterranean), 55, 129
NATO and North Atlantic Treaty, 52–53, 57, 65
NATO and out-of-area issue, 50–51, 54, 57–65, 66, 121–122, 123–124, 128–131
NATO and PfP (Partnership for Peace), 128, 143, 182, 200, 228
and Mediterranean, 167, 180, 181, 184–185, 217, 228–229
NATO and Political Committee: and NATO's Mediterranean Initiative, 172, 179
NATO and STANAVFORMED (Standing Naval Force Mediterranean), 55, 129
and NATO's Mediterranean Initiative, 181
NATO and Strategic Concept (1991), xii-xiii, 129, 130, 140–142, 151, 155, 156, 159
NATO and Strategic Concept (1999), xii-xiii
NDP (National Democratic Party), 77
Netanyahu, Binyamin, xi, xii, 14, 15, 16, 72
Netherlands, 59
Mediterranean, relations with, 59
NATO, relations with, 52, 53
problem of migration with, 99
North Korea, 109
NPT (Nuclear Non-Proliferation Treaty), 110, 127, 156

OAU (Organization of African Unity), 111, 200
OCME (Organization for Cooperation in the Middle East), 208
Operation Desert Storm, 12, 100, 121–123, 128
Ordonez, Francisco Fernandez, 134
OSCE (Organization for Security and Cooperation in Europe), 41–42,

72, 87 (*see also* CBM, CSCE, EUROFOR, EUROMARFOR)
OSCE-Mediterranean Dialogue, 8, 31, 147–148, 205–208, 225 (*see also* Barcelona Process, CBM, CSBM, CSCE, WEU-Mediterranean Dialogue)
  Algeria, relations with, 208
  Egypt, relations with, 147, 205–206
  Israel, relations with, 206, 207
  Italy, relations with, 147
  Jordan, relations with, 205
  Malta, relations with, 147
  Mauritania, relations with, 205, 208
  NATO's Mediterranean Initiative, relations with, 116, 197, 205, 208, 217–218
Ould Daddah, Ahmed, 75, 78, 86
Ould Taya, Maaouya, 75, 86

Pakistan, 3, 136
Perry, William, 139, 157
PI (Istiqlal Party), 78
PLO (Palestine Liberation Organization), xi, 13, 15, 62
Poland
  NATO, relations with, 228
POLISARIO (Popular Front for the Liberation of Saguia el-Hamra and Rio de Oro), 11–12, 110
Portugal, 49, 71, 107, 115, 151 (*see also* CSCM, EU, Mediterranean Forum, U.S., Western Mediterranean Forum, WEU, WEU-Mediterranean Dialogue)
  Algeria, relations with, 104, 115, 132, 171
  Libya, relations with, 132
  Mauritania, relations with, 122–123, 169, 170
  Mediterranean, relations with, 2, 63, 131, 132, 137, 168
  Morocco, relations with, 132
  NATO, relations with, 52, 53, 56
  NATO's Mediterranean Initiative, relations with, 155, 169, 170, 171, 183, 184–185, 187
  problem of migration with, 10–11, 102
  Tunisia, relations with, 115
PRDS (Republican Democratic and Social Party), 78, 84
Preventive diplomacy
  and Barcelona Process, 214, 217
  definition of, 17–18
  and NATO, 129, 141, 142, 143, 159
  and NATO's Mediterranean Initiative, 2, 19, 20, 45, 115, 192, 231
  and UN, 17

Quadrangolare, 132

Rabin, Yitzhak, 13, 14
RAND Corporation, and NATO's Mediterranean Initiative, 175, 180, 188, 192, 213, 226, 230–231
RCD (Constitutional Democratic Rally) (Tunisia), 75, 77–78, 82
RCD (Rally for Culture and Democracy) (Algeria), 79, 90
RDF (Rapid Deployment Force), 59–60
Reagan, Ronald, 148
REDWG (Working Group on Regional Economic Development), 125, 128, 151, 210
Region, definition of, 5–9
Rifkind, Malcolm, 208
RND (National Democratic Rally), 79

Romania, 186
Russia (*see also* CSCM, Mediterranean Forum)
  Iran, relations with, 109
  Mediterranean, relations with, 49
  NATO, relations with, xiii, 2, 66, 143, 184

Sahrawi Arab Democratic Republic, 110
Samaras, Antonis, 134
Saudi Arabia, 92, 93, 123, 127
Security
  and cultural differences, 28–31, 113–115
  definition of, 23–28
Security community, definition of, 43–44
Security regime, definition of, 43
SFOR (Stabilization Force), 4, 124, 174, 190
Shas party, xii
Solana, Javier, xiv, 180, 182, 183, 185, 189, 192, 194n, 197–198, 199, 217, 230
South Korea, 206
Soviet Union, 13, 87, 122 (*see also* Turkey, U.S.)
  Afghanistan, relations with, 55, 59
  Algeria, relations with, 51, 54
  Egypt, relations with, 51, 54
  Libya, relations with, 51, 52, 54
  Mediterranean, relations with, 49, 50, 51–52, 54–56, 61, 65–66, 135
  NATO, relations with, 49. 50. 51, 52, 56, 57–59, 61, 121, 140, 141
  Syria, relations with, 51, 54
Spain, 49, 71, 107, 115, 151 (*see also* CSCE, CSCM, EU, Mediterranean Forum, U.S., Western Mediterranean Forum, WEU, WEU-Mediterranean Dialogue)
  Algeria, relations with, 104, 115, 132, 171
  Libya, relations with, 132
  Mauritania, relations with, 63–64, 122–123, 169, 170
  Mediterranean, relations with, 2, 52, 63, 130, 131, 132, 137, 149, 155, 168, 183
  Morocco, relations with, 63–64, 101, 117n, 132
  NATO, relations with, 51, 56, 158, 183, 203, 223
  NATO's Mediterranean Initiative, relations with, 155, 169, 170, 171, 172, 187
  problem of migration with, 10–11, 79, 99, 100, 101
  Tunisia, relations with, 63–64, 115
Suez Crisis, 54, 57
Syria, 13, 65–66, 125, 126, 127, 131 (*see also* Barcelona Process, CSCE, EU, Jordan, Mediterranean Forum, Morocco, NAA-Mediterranean Dialogue, Soviet Union, Turkey)
  EC, relations with, 151
  military capability of, 109
  NATO's Mediterranean Initiative, relations with, 169, 171, 190, 226, 228

Tanir, Vefa, 186
Tunisia, 107, 114, 124, 137 (*see also* CSCE, EU, France, Italy, Mauritania, Mediterranean Forum, NAA-Mediterranean Dialogue, Portugal, Spain, US, Western Mediterranean Forum, WEU-Mediterranean Dialogue)
  Algeria, relations with, 82, 102, 115, 229
  AMU, relations with, 107, 112, 133, 150

economic situation in, 90, 98, 106
EC, relations with, 107, 151
EUROFOR, relations with, 204
Libya, relations with, 63, 112, 173
NATO's Mediterranean Initiative, relations with, 1, 85, 157, 169, 178, 190, 229
political situation in, 74–75, 77–78, 81, 82, 83, 85, 90, 91
problem of migration with, 98–99
role of military in, 87
Tunisian Islamic Front, 82
Turkey, 9, 55, 86, 88 (*see also* Barcelona Press, EU, Mediterranean Forum, U.S.)
  Greece, relations with, 10, 18, 52, 55–56, 132–133, 223, 225, 227
  Iran, relations with, 2, 10, 64, 109, 121, 186
  Iraq, relations with, 2, 10, 64, 121, 122, 133, 186
  Israel, relations with, 186
  NATO, relations with, 51–52, 53–54, 55–56, 123, 186
  NATO's Mediterranean Initiative, relations with, 185–186, 223
  Soviet Union, relations with, 50, 52
  Syria, relations with, 2, 10, 64, 109, 121, 186

UC (Constitutional Union), 78
UDF (Union of Democratic Forces), 78, 84
Ukraine
  NATO, relations with, xiii, 143
UN (United Nations), 200 (*see also* EUROFOR, EUROMARFOR, WEU)
  Iraq, relations with, 11, 90, 105, 109
  Israel, relations with, 11, 14, 16
  Libya, relations with, 11, 112, 113, 133, 137, 215
  NATO, relations with, 122, 124, 174
  Western Sahara, issue of, 110–111
United Arab Emirates, 208
UNPROFOR (United Nations Protection Forces), 124
U.S. (United States), 2, 11, 59–60, 121–122, 131, 132, 148 (*see also* Barcelona Process, CSCM, Middle East peace process)
  Algeria, relations with, 64, 102, 171
  Britain, relations with, 58, 59, 60
  Egypt, relations with, 64, 127
  France, relations with, 53, 56, 58, 59, 60, 102, 143, 186–187, 223
  Israel, relations with, 2, 14–17, 51, 61, 62, 66, 136, 169–170, 186, 191
  Italy, relations with, 56, 59, 60, 62
  Jordan, relations with, 105
  Lebanon, relations with, 59
  Libya, relations with, 5, 10, 11, 60, 62, 136
  Mediterranean, relations with, 9, 50, 52, 54, 58–60, 110, 132, 156, 168, 186–187, 211, 223
  Morocco, relations with, 64
  NATO, relations with, 9, 52–53, 54, 64, 143, 157, 186–187
  NATO's Mediterranean Initiative, relations with, 155, 169–170, 186, 231
  Portugal, relations with, 61
  Soviet Union, relations with, 55–56, 58–59, 61, 64, 122
  Spain, relations with, 56
  Sudan, relations with, 11

U.S. (United States) (*continued*)
  Tunisia, relations with, 64
  Turkey, relations with, 55–56
USFP (Socialist Union of Popular Forces), 78

Virtue party, 88

Wafd party, 77
Welfare party, 88
Western Mediterranean Conference, 137–138
Western Mediterranean Forum, 50, 137–139, 144–145
  Algeria, relations with, 50, 138, 139
  AMU, relations with, 138, 139, 144
  Egypt, relations with, 139, 144–145
  France, relations with, 50, 135, 137–139, 145
  Greece, relations with, 139
  Italy, relations with, 50, 138–139
  Libya, relations with, 50, 138, 139
  Malta, relations with, 50, 138
  Mauritania, relations with, 50, 138
  Morocco, relations with, 50, 138
  Portugal, relations with, 50, 135, 138
  Spain, relations with, 50, 138
  Tunisia, relations with, 50, 138
Western Mediterranean Pact, 52
Western Sahara, 11–12, 20, 63, 86, 92, 110–111, 112, 138 (*see also* UN)
WEU (Western European Union), 80–81, 114–115, 148–151, 200 (*see also* EU, EUROFOR, EUROMARFOR)
  Bosnia, relations with, 102, 149
  France, relations with, 132, 202
  Greece, relations with, 149–150
  Italy, relations with, 202, 203
  Mediterranean, relations with, 149–150
  NATO, relations with, 124, 143, 149, 201–202
  Portugal, relations with, 149–150
  Spain, relations with, 149–150, 202, 203
  UN, relations with, 124, 149
WEU-Mediterranean Dialogue, xiv, 8, 148, 150–151, 159, 171, 192, 197, 199–204 (*see also* Barcelona Process)
  Algeria, relations with, 150, 200, 202
  CSCE, relations with, 150
  Cyprus, relations with, 150
  Egypt, relations with, 150, 189–190, 200
  France, relations with, 200
  Greece, relations with, 199
  Israel, relations with, 150, 199, 200
  Italy, relations with, 200
  Jordan, relations with, 199
  Libya, relations with, 150
  Mauritania, relations with, 150, 200
  Morocco, relations with, 150, 200
  NATO's Mediterranean Initiative, relations with, 116, 197, 202–204, 217
  OSCE, relations with, 150, 151
  Portugal, relations with, 200
  Spain, relations with, 200
  Tunisia, relations with, 150, 200
  Turkey, relations with, 200
WMD (Weapons of mass destruction), 11, 14, 40, 108–110, 126, 127, 159 (*see also* Barcelona Process, Mediterranean Forum)
  and NATO, xiii, xiv, 10, 108, 110, 121, 123, 128, 130–131, 140, 156

and NATO's Mediterranean Initiative, 130–131, 174–175, 180, 189, 226, 230
World Bank, 98, 106
Wörner, Manfred, 123, 128, 141

Yahia, Habib Ben, 101, 113–114, 204

Yemen, 55
Yugoslavia, 132 (*see also* CSCE, CSCM)
    NATO, relations with, 10, 55, 122, 123–125, 128, 129

Zeroual, Liamine, 75–76, 81, 86